Patrick Moore

FIRESIDE ASTRONOMY

Galileo. In this artist's impression, Galileo is passing
the volcanic satellite Io on its way into the first orbit
round Jupiter. Whether or not Galileo will be a
success remains to be seen; we can only hope.
Reproduced by kind permission of NASA.

Patrick Moore

FIRESIDE ASTRONOMY

An anecdotal tour through the history and lore of astronomy

JOHN WILEY & SONS

Chichester · New York · Brisbane · Toronto · Singapore

Wiley Editorial Offices

John Wiley & Sons Ltd
Baffins Lane, Chichester
West Sussex PO19 1UD, England

John Wiley & Sons, Inc., 605 Third Avenue
New York, NY 10158-0012, USA

Jacaranda Wiley Ltd, G.P.O. Box 859, Brisbane
Queensland 4001, Australia

John Wiley & Sons (Canada) Ltd, 22 Worcester Road,
Rexdale, Ontario M9W 1L1, Canada

John Wiley & Sons (SEA) Pte Ltd, 37 Jalan Pemimpin #05-04
Block B, Union Building, Singapore 2057

British Library Cataloguing in Publication Data

A catalogue record for this book is available from the British Library

ISBN 0 471 93164 0

Typeset in 10½/12 pt Plantin by Dobbie Typesetting Limited, Tavistock, Devon.
Text design and text format by Russell Townsend
Cover design by David Chandler, Graphic Designer, John Wiley
Printed and bound in Great Britain by Courier International Ltd., East Kilbride, Scotland.

Preface

Many books on astronomy have been published in recent years. Almost all of them are written to a set plan: either they are textbooks, general surveys, or else devoted to some particular branch of astronomical science. The present offering is quite different. It is most emphatically not a textbook, and, like its predecessor (*Armchair Astronomy*) it has no set plan at all. Please just 'dip into it'!

What I have tried to do is to entertain you with facts and episodes which you may not have encountered before. For example, did you know that there was once an African republic named after a comet; that a full-scale space search was once sparked off by a mousetrap; that a prize was offered less than a century ago to be given to the first man to establish contact with beings from another world—excluding Mars, which was too easy; or that an American professor plans to improve our weather by blowing up the Moon? I can assure you that all this is true. Just in case the book falls into the hands of a reader who does not know an asteroid from an adenoid, I have added a brief Glossary—but even without this, I hope that you will be able to dip into the very short chapters at random, and enjoy them. I have done my best.

PATRICK MOORE

Selsey
September 1991

Contents

A Brief Glossary

Asteroids (or minor planets) Small members of the Solar System, most of which (though not all) keep to the region between the orbits of Mars and Jupiter.

Astronomical unit The mean distance between the Earth and the Sun; in round figures, 93,000,000 miles.

Aurora Luminous glow in the Earth's upper air, caused by charged particles from the Sun; Aurora Borealis in the northern hemisphere, Aurora Australis in the southern.

Binary star system A star system made up of two components, genuinely associated and moving round their common centre of gravity.

Black hole A region round an old, collapsed star which is now pulling so strongly that not even light can escape from it.

Comet A member of the Solar System; it has an icy nucleus, and when the comet nears the Sun the ices begin to evaporate, producing a head or coma and (sometimes) a tail or tails.

Cosmic rays Not true rays, but high-velocity atomic particles coming from space in all directions.

Eclipse, lunar The passage of the Moon through the cone of shadow cast by the Earth.

Eclipse, solar The temporary blotting-out of the Sun by the Moon, when the Moon passes between the Sun and the Earth.

Electromagnetic spectrum The whole range of radiation wavelengths from the very long radio waves to the ultra-short gamma-rays.

Electron A fundamental particle with unit negative electrical charge.

Escape velocity The minimum velocity which an object must be given if it is to escape from the pull of another body, assuming that it is given no extra impetus.

Galaxies Systems made up of stars, nebulæ and interstellar matter. Some, though by no means all, are spiral in shape.

Galaxy, the The Galaxy of which our Sun is a member. It contains about 100,000 million stars.

Light-year The distance travelled by light in one year: 5.88 million million miles.

Meteor A small particle, of sand-grain size, which dashes into the upper atmosphere and burns away. Meteors are cometary débris.

Meteorite A larger body which may land on Earth intact, and even produce a crater. A meteorite comes from the asteroid belt, and is not associated with comets or meteors.

Milky Way The luminous band in the night sky seen when we look along the main plane of the Galaxy.

Molecule An atom group, such as H_2O or water (two atoms of hydrogen, one of oxygen).

Nebula A cloud of gas and dust in space, inside which fresh stars are being produced.

Neutrino A particle with no electrical charge, and (probably) no mass.

Neutron A fundamental particle with no electrical charge. A proton and an electron may combine to make a neutron.

Neutron star The remnant of a supernova outburst—a very small super-dense object, made up of neutrons and in rapid rotation.

Nova The sudden, temporary flare-up of the White Dwarf component of a binary system.

Occultation The covering-up of one body by another. (To be strictly accurate, a solar eclipse is not an eclipse at all, but an occultation of the Sun by the Moon.)

Orbit The path of a celestial body in space.

Planet A non-luminous body in orbit round a star.

Prominences Masses of glowing hydrogen rising from the surface of the Sun. With the naked eye they can be seen only during a total solar eclipse.

Proper motion, stellar The individual movement in the sky of a star against the background of more distant stars.

Proton A fundamental particle with unit positive electrical charge.

Pulsar A rotating neutron star which is sending out beams of radio radiation from its magnetic poles.

Satellite A secondary body moving round a planet.

Shooting-star The luminous effect seen when a meteor burns away in the upper atmosphere.

Sidereal period The time taken for a planet or comet to complete one orbit round the Sun, or for a satellite to complete one orbit round its primary.

Solar wind A stream of atomic particles being sent out constantly by the Sun.

Star A globe of gas, shining by internal nuclear reactions in the case of a normal star such as the Sun.

Supernova A violent stellar outburst, due either to the total destruction of a White Dwarf (Type I) or the collapse of a very massive supergiant star (Type II).

White Dwarf A very old, collapsed star, which has used up all its nuclear energy.

Zodiac A belt round the sky, extending to 8 degrees on either side of the ecliptic, in which the Sun, Moon and principal planets are always to be found.

A Recipe from Copernicus

There can be little doubt that the Polish astronomer Mikołaj Kopernik—better remembered as Copernicus—is one of the great figures in scientific history. It was he who revived the old idea that the Earth moves round the Sun instead of vice versa, and his book published in 1543 really marks the start of what we may loosely call the modern era. Predictably, the official Church was not impressed, and that enlightened reformer Martin Luther commented that 'this fool seeks to overturn the whole art of astronomy', but Copernicus himself escaped prosecution by the simple act of dying. For many years he had been installed at Frombork in the province of Warmia, which he himself described as 'the remotest corner of the earth', and he had been sensible enough to withhold publication until the very last days of his life.

As a young student he attended the University of Cracow, one of the most famous in Europe, and it was here that he began medical studies, though this was not the main reason for his going there. Some of the traditions would seem rather out of place today. For example, when a public execution was due to take place in the Market Square, the whole of the medical faculty, including the Dean, would don official robes and attend the ceremony, taking careful notes while the hapless victim was drawn and quartered. When there were no convenient executions, the faculty would go instead to the local slaughterhouse—presumably without bothering to put on their robes—and use pigs as substitutes. Whether Copernicus attended any of these festivities is not known, but presumably he did.

His next move was to Padua University, in 1501, where he spent some time in medical studies, taking a full three-year course. This was quite in order. He was already a Church official, and it had always been intended that he should return to Poland when his training was complete, mainly to act as assistant to his uncle—a senior Bishop in Warmia—but also to double as what we would now call a GP. Some of his notebooks survive, and in the margins are some interesting handwritten recipes. For example:

Copernicus—more properly Mikołaj Kopernik.

'Take resin from a fruit tree, boil it in beer three times, and drink with meals: helps podagra.' (Podagra, in case you didn't know, is gout in the feet.)

'A remedy against paralysis of the body: make an infusion of sage leaves, add rue, castoretum, boil in wine and drink.'

'A note on corns—look in the Pandects about the weeping willow.'

'For a remedy to protect one against the bite of a rabid dog, see the Pandects under the word *sapphire*.'

'Water extracted from beech leaves helps everything.'

And if you want a kind of universal remedy:

'Take two ounces of Armenian clay, a half ounce of cinnamon, two drachms of tormentil root, dittany, red sandalwood, a drachm of ivory and iron shavings, two scruples of ash and rust, one drachm each of lemon peel and pearls; add one scruple each of emerald, red hyacinth and sapphire; one drachm of bone from a deer's heart; sea locusts, horn of a unicorn, red coral, gold and silver foil—all one scruple each; then add half a pound of sugar, or the quantity which one usually buys for one Hungarian ducat's worth.' Significantly, he added: 'God willing, it will help.'

I would be frankly unwilling to sample this concoction, particularly as Copernicus added the cheerful rider that 'Those who inherit diseases are rarely cured of them, and will be wise to endure their suffering in patience'. Yet when he was back in Warmia he became widely known as a medical expert, and it is said that during a serious epidemic in 1519 his skill saved many lives.

By now he was working hard to perfect his theory of the universe, but he was also pressed into service to help in the defence of his country against the Teutonic Knights, an unpleasant religious-cum-military organization with the official title of the Order of the Hospital of the Holy Virgin Mary of the German House in Jerusalem. The Order had been founded during the Crusades, but had become an unmitigated nuisance, and it was not until 1525 that it finally withdrew from Warmia. It was eventually suppressed in the early nineteenth century by Napoleon Bonaparte. (Incidentally, its members were in general recruited from German knights of noble lineage; the accepted dress was a white cloak decorated with black crosses—forerunners of the Ku Klux Klan?)

As a personality, Copernicus seems to have been lacking in charisma. His biographer, the French astronomer Pierre Gassendi, wrote of him that late in his career 'he felt revulsion toward any form of familiarity and any frivolous and pointless conversation', so that he would have been disinclined to sample the night life of Frombork even if there had been any. At least he made his mark in history, and his medical recipes are worth remembering, despite the sea locusts and the bone from a deer's heart!

The Strangest Planet

The 21st General Assembly of the International Astronomical Union, held in Buenos Aires in July 1991, was dominated by the spectacular announcement that a new planet had been discovered—not in our Solar System, but moving in orbit round a neutron star 30,000 light-years away, not very far from the centre of the Galaxy. I first heard about it on the morning of Wednesday 24th,

The Lovell Telescope at Jodrell Bank. The dish is 250 feet in diameter.

from Professor Sir Francis Graham-Smith, who came into the editorial office with a paper which he wanted to use in the following day's issue of the IAU newspaper. (Needless to say, I made sure that it took up the whole of the front page.) To say that I was taken aback is a gross understatement!

Of course, the object could not be seen, and we had to depend upon observations at radio wavelengths. The research was done by Setnam Shemar, a member of the team at Jodrell Bank in Cheshire which was led by Professor Andrew Lyne (the other member of the trio was Dr Matthew Bailes). To explain how they worked, I must digress for a few moments to say something about neutron stars in general. Considerable numbers of them have now been located.

A neutron star is a curious thing. It is the remnant of a very old, massive star which has exploded as a supernova, blowing most of its material away into space and leaving only its core, which is very small, incredibly dense and in rapid rotation. As a neutron star spins, beams of radio radiation from its magnetic poles sweep across us in the manner of a lighthouse, and we pick up the pulsed emissions—which is why neutron stars are known as pulsars.

In 1985 a survey from Jodrell Bank resulted in the discovery of forty new pulsars. One of them, PSR 1829-10, behaved in an unusual way. It span round 330 times per second, but there were variations which appeared to be erratic. Shemar made some more precise timings, and concluded that the variations were periodic. This led to the conclusion that the neutron star was being pulled on gravitationally by an orbiting body about ten times as massive as the Earth, moving in an almost circular orbit at a distance of about 66,000,000 miles from the star—about the same as the distance between Venus and the Sun. The period of revolution was given as 184 days or six months, and it was calculated from the present-day behaviour of the pulsar that its age should be of the order of one and a quarter million years.

A body only ten times the mass of the Earth could not be a star, but whether it could be classified as a planet was another matter. But was there an orbiting body at all? From the outset I admitted to having doubts, mainly because the stated orbital period of the 'planet' was six months, and it seemed at least possible that the observations simply reflected our own movement round the Sun. All in all, it seemed wise to be very cautious before claiming that the first extra-solar planet had definitely been found.

Remember that a supernova outburst is the most violent phenomenon in the whole of nature, and would destroy any planet within range; at its peak, a supernova can shine as powerfully as all the stars in an average galaxy combined. No planet within reasonable range could survive an outburst of that sort. So if a planet existed, we had to assume either that the pulsar PSR 1829-10 was formed in a completely different way, not involving a supernova outburst, or else that the planet was not in orbit when the flare-up happened. Both of these explanations seemed wildly unlikely.

The mystery was solved in January 1992. There had indeed been a major mistake. The calculations had been worked out on the assumption that the Earth's orbit round the Sun is circular, but of course it is not—it is slightly but definitely elliptical (our distance from the Sun ranges between 91½ million miles in December out to 94½ million miles in July). When the Jodrell Bank astronomers re-processed their data, taking the ellipticity of the Earth's orbit into account, they found that everything fell into place. There were no unexplained movements, and no orbiting planet. In an official statement released on January 15 1992, the team wrote that 'Our failure to recognize the result of a position difference and to perform the usual procedure has resulted in the apparent planet, and we must accept full responsibility for this error'.

It shows that astronomers are human enough to make mistakes—and it was greatly to the credit of the Jodrell Bank researchers that they made a full statement as soon as they realized what had happened. It was a pity that the planet proved to be non-existent. If it had really been there, it would have been the strangest world imaginable.

Interestingly, planets orbiting another pulsar—PSR 1257 + 12—were reported by American astronomers just before the Jodrell Bank announcement in January 1992. Whether or not these exist must be a matter for debate, but I am frankly sceptical, and it seems to me much more likely that there have been errors of the same kind. It is all most disappointing!

☾3 Down Among the Penguins

Would you like to go to the South Pole? I admit that I would be fascinated (the furthest south I have ever managed, so far, is Invercargill in New Zealand). Astronomically, too, the South Pole has some advantages, and it is a great deal more accessible now than it was when Amundsen and Scott visited it less than a century ago.

The whole question was discussed at a special session of the International Astronomical Union, held at Buenos Aires in 1991, when the idea of a major

The Southern Lights. Old engraving of the Aurora Australis.

observatory either at the Pole or very close to it was considered in detail. Of course, there have been scientific bases in Antarctica for years now, but the present scheme is much more ambitious, and would involve setting up very large telescopes.

Antarctica is suitable for various reasons. First, you have an uninterrupted view of the sky during the six-months' night, and there is a distinct chance that some important event might be observable from a dark-sky polar site and nowhere else. The atmosphere is also very dry and transparent, and there is not a great deal of cloud. The dryness is particularly attractive to astronomers who specialize in infra-red work, because atmospheric water-vapour is the enemy of the infra-red worker, and there is not much of it over the South Pole.

Moreover, the popular view that Antarctica is a permanently windswept, icy country, deep in snow and left entirely to the penguins and the seals (no polar bears, fortunately!) is not entirely correct. Parts of it are high; the eastern plateau goes up to 13,000 feet, and there is very little snowfall. Neither are there violent gales, because the wind velocity seldom exceeds 10 feet per second. The main problem, obviously, is the cold. Vostok station, the highest manned base at present—some way from the actual Pole—experiences temperatures which go down to more than 80 degrees below zero Centigrade in winter, and even during the short summer the climate is never what may be described as balmy. Also, the temperature gradient is quite sharp from ground level, and so it is a good idea to raise your main telescope as high as possible. Yet it is quite likely that seeing conditions in Antarctica may turn out to be better than anywhere else in the world, even the top of Mauna Kea in Hawaii.

Investigations have already been started in earnest. In 1991 came the establishment of the Centre for Astrophysical Research in Antarctica (CARA), which is all-American at the moment but will surely become international in the near future. And at the 1991 meeting of the International Astronomical Union, a resolution was passed calling for national committees on astronomy to work together with national Antarctic agencies in planning a full-scale observatory on the high plateau—possibly at Dome C, beyond Vostok Base if you start from the Pole.

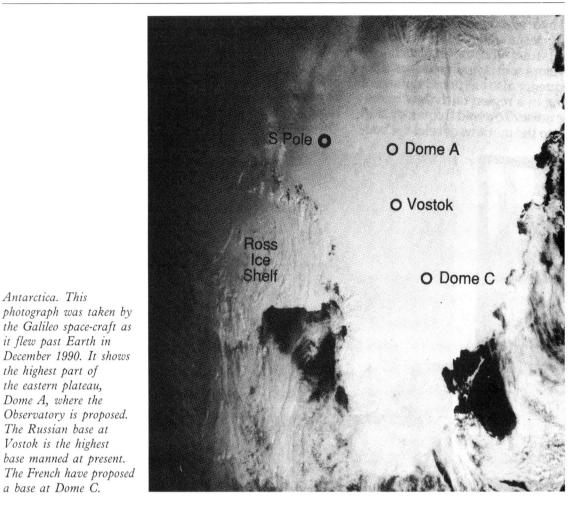

Antarctica. This photograph was taken by the Galileo space-craft as it flew past Earth in December 1990. It shows the highest part of the eastern plateau, Dome A, where the Observatory is proposed. The Russian base at Vostok is the highest base manned at present. The French have proposed a base at Dome C.

I hope it happens. Think, too, of the wonderful displays of Auroræ Australis or Southern Lights during the winter—and the joy of a six-months' day. If you become bored, of course, you can always go out and try to tame a penguin or two—bearing in mind that you will have to put on some warm clothing!

☾4 The Guzman Prize

The name of Camille Flammarion is known to all astronomers. He was French, and founded the prestigious journal *L'Astronomie*; he worked for a time at the Paris Observatory, and then founded his own observatory at Juvisy; he was a skilled observer, particularly of the planets, and he was also an energetic popularizer. I never knew him, because he died in 1923 (the year I was born), but I did know his second wife, née Renaudot, who was still alive in the 1950s.

Flammarion wrote two vast books about the history of the observation of Mars. They are quite fascinating. I have translated them from the French (at the request of the Lowell Observatory) and a few manuscript copies exist. One remarkable section, dated 1899, refers to the Guzman Prize. I doubt whether many people have ever heard of it, and so I propose to quote exactly what Flammarion wrote (my translation, of course). The section runs as follows:

'*Communications with Mars.*—One of my readers, or, rather, the mother of one of my enthusiastic readers—Mme Guzman of Bordeaux—has had the generous idea of presenting, in memory of her son, a prize of 100,000 francs, left to the Academy of Sciences to be given to anyone who can find a method of communication between the Earth and another world.

'Unfortunately M. Guzman followed all my writings about Mars so avidly that he has commented that mankind need wait no longer before

Martian deserts, as seen from Viking 1. The great canyon Valles Marineris cuts across the lower centre of the mosaic; to the right the volcanoes Arsia Mons, Pavonis Mons and Ascræus Mons. The red deserts are splendidly shown—and this is where Charles Cros proposed to burn messages! Reproduced by kind permission of NASA.

communicating with the Martians, since the problem has already been presented and half solved; and his mother has therefore *excluded* the Martians from this magnificent competition!

'Leaving out the only planet which holds out any possibilities is surely original. Terrestrial humanity must be truly astounded.

'Perhaps the founder wishes to immortalize her son's name *for eternity*.

'The Academy accepted the Prize, not without hesitation and not without various discussions. It will commemorate the name of Guzman for a long time, because if we are ever to establish communication it will probably be with Mars. Happily, the interest of the capital can be used for encouraging the progress of astronomy.

'On 17 December 1990 the Academy Published an announcement of the prize, in the following terms:

PIERRE GUZMAN PRIZE

Mme Clara Gouguet, née Guzman, has left to the Academy of Sciences a sum of 5000 francs for the foundation of a prize which will carry the name of the Pierre Guzman Prize, in memory of her son, to be presented to the person who first finds a method of communicating with another world, excluding the planet Mars.

Assuming that the prize of 5000 francs will not be distributed at once, the founder wishes that until the prize is won, the interest on the capital accumulating for five years will provide a prize, also given the name of Pierre Guzman, for the French or foreign scientist making the most important progress in astronomy.

The quinquennial prize, represented by the interest on the capital, will be given for the first time—if it remains—in 1905.

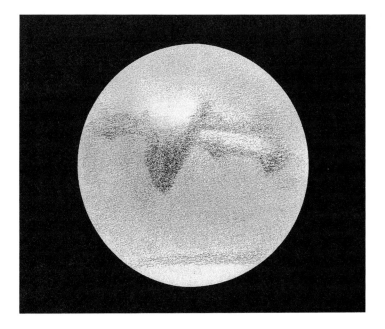

Mars, as I drew it with my 15-in reflector on 5 November 1990. The triangular Syrtis Major is clear.

The Prize was in fact awarded in 1905, so Flammarion tells us, jointly to M. Perrotin, Director of the Nice Observatory (who had died suddenly in the previous year) and to M. Fabry, one of the astronomers at the Marseilles Observatory. To the best of my knowledge, they were the first and only recipients. What has happened to the accumulated interest since that time I know not!

Of course, the idea of calling up the Martians is far from new. During the last century there were serious suggestions about drawing large patterns in places such as the Sahara Desert to attract the attention of watchers on the Red Planet, and in the 1870s an enthusiastic Frenchman, Charles Cros, tried to interest the Government in a plan for building a large burning-glass and focusing the beam on the Martian deserts, swinging it about and writing words. I have always wondered what words he hoped to write—and I am also curious about our own reactions if a message suddenly began to appear in our own Sahara. I fear it is not likely to happen. But I cannot resist a final quote, coming from a 1924 issue of the periodical *Wireless Weekly*. The article, written by a G. V. Dowling, describes 'the world's largest wireless receiver', with which members of his staff proposed to listen in to messages from Mars. What he called the 'P. W. 24-valve receiver' was 'a set comprising 20 stages of HF amplification, detector, and three stages of note magnification'. He went on as follows: 'For the purpose of endeavouring to receive messages from the planet Mars, a special 12-foot frame aerial will be used, together with certain coils which are the subject of letters patent'. He proposed to listen out at 'extremely long wavelengths, as we are led to believe that success, if it should come to us, will be achieved by this means.'

It wasn't!

C5 The Hubble Telescope: Failure or Success?

In April 1990 the Space Shuttle launched the world's most futuristic telescope: the HST or Hubble Space Telescope, named in honour of the man who first proved that the dim, misty patches which we used to call spiral nebulæ are in fact true galaxies, so remote that their light takes millions, hundreds of millions or even thousands of millions of years to reach us. The HST has a mirror 94 inches across, and has been put into an orbit which takes it round the world every 95 minutes at an altitude of between 300 and 400 miles.

In space, above the atmosphere, seeing conditions are perfect all the time, and it was confidently expected that the Hubble Telescope would far out-perform any instrument based at ground level. Too late, it was found that the mirror had been wrongly shaped. There is a major error in the optics, and this results in what astronomers call spherical aberration, so that star images are surrounded by 'fuzz' instead of being sharp and clear-cut. Moreover the solar panels, which collect the Sun's light and convert it to power for the

Jupiter, seen by the Hubble Space Telescope on 11 March 1991; the Planetary Camera was used (green light), The Great Red Spot is to the lower right; also on the right, near Jupiter's equator, the satellite Europa is about to be occulted by the limb of the planet. The J-shaped clouds along the equator indicate intense jet-streams in the Jovian atmosphere. The picture is equivalent to Voyager images taken 5 days before closest approach in 1979.

instruments, are also faulty; they flap. Finally the gyroscopes, which stabilize the entire telescope, are giving constant trouble. All in all, these various problems have led many people to dismiss the telescope as an expensive failure.

This is completely wrong. True, the telescope is not as perfect as it ought to have been, but although some of its planned programmes have had to be abandoned it can still produce results which no other telescope can hope to match. Look, for example, at the pictures of Mars, Jupiter and Saturn; one might almost think that they had been taken by a passing space-craft. Remote Pluto is shown as distinctly separate from its companion Charon; previously all we had been able to do was to obtain a merged image. Hubble can look into the heart of the Orion Nebula, where fresh stars are being created, and into the cores of dense globular clusters. It can study the surroundings of the supernova which flared up in the Large Cloud of Magellan in 1987, and it has given us a superb view of a 'gravitational lens' effect, where we see multiple images of a distant quasar whose light has been bent by an intervening galaxy. Moreover, the HST can operate at ultra-violet wavelengths.

So do not think solely about the imperfections in the mirror, regrettable though they are. When history comes to be written, the Hubble Space Telescope will be remembered not because of its faults, but as the instrument which ushered in a whole new era of scientific research.

Planets of Other Stars

6

Of all the questions that I am asked, I suppose that the favourite is: Can there be life elsewhere? I always have to admit that although I believe that there must be plenty of life in the universe, I cannot prove it. I cannot even be absolutely sure that there are planets moving around other stars. It seems probable; after all, our Sun is only one of 100,000 million in our Galaxy alone, and it is very ordinary. Unfortunately, we have yet to see planets of other stars, because a planet is so very small compared with a normal star, and has no light of its own.

Up to now, astronomers who are hunting for other planetary systems have tried various methods. One depends on the pull of gravity. The stars are moving around in space, and although their individual motions seem very slow (which is why the constellation patterns look the same now as they did in the time of Julius Cæsar) they can be measured. A massive planet moving round a relatively lightweight star would make the star 'wobble' very slightly, and this has been suspected with several dim, close stars, but the evidence is not yet conclusive. A second method has been proposed recently by two astronomers at Princeton University in America, Shude Mao and Bohdan Paczyński, who believe that a large planet passing in front of a normal star would dim the star's light just sufficiently for us to be able to notice it. Whether this is practicable remains to be seen. So what about the chances of actually seeing an extra-solar planet as a speck of light?

The one telescope which might be regarded as capable of doing so is the Hubble Space Telescope. Sadly, the fault in the main mirror means that this particular hunt has had to be given up, at least until (or if) repairs can be carried out. For the moment, then, our best hopes rest with infra-red methods, and this brings me on to the story of Beta Pictoris.

To the naked eye Beta Pictoris is an unremarkable star. Its apparent magnitude is 3.8; it is 40 times as luminous as the Sun, and is 78 light-years away. It is white, so that its surface is rather hotter than that of our yellow Sun. It is too far south to be seen from anywhere in Europe or the United States; it lies in the little constellation of the Painter, not far from the brilliant Canopus. Until 1983 nobody paid any particular attention to it.

In that year, however, the Infra-Red Astronomical Satellite IRAS was launched. Even though it was operational for less than a year, it ranks as one of the most successful satellites in the history of space research. When the investigators—Drs Gillett and Aumann—turned it toward some stars, such as Vega in Lyra, IRAS recorded what is called an infra-red excess—that is to say radiation from material which is not hot enough to shine at optical wavelengths, but emits in the infra-red (just as an electric fire will do after you have switched it on, before the bars begin to glow). It was suggested that this material might be planet-forming, or could even indicate a fully-fledged planetary system.

Another star to show a large infra-red excess was Beta Pictoris. Two eminent observers, Bradford Smith and Richard Terrile, decided to find out whether anything could be seen in visible light, so they went to the Las Campanas

Infra-red investigators: Drs Gillett and Aumann, who made the first IRAS studies of 'infra-red excesses' round other stars (1983).

Observatory in Chile and used the powerful telescope there together with very sensitive electronic equipment. Sure enough, they detected material extending to either side of the star, out to a distance of 400 times the distance between the Earth and the Sun, with the thickest part within 100 times the Earth–Sun distance. In fact they were seeing a disk of 'dust' almost edgewise-on.

The star itself is dimmed by only half a magnitude. This indicates that the dust does not extend all the way to the star's surface, and it is in this region that planets would probably have been formed. Very small particles would have been swept in, leaving only the larger ones; Richard Terrile commented that 'it would be hard *not* to form planets from material like this'.

Confirmation that the disk is due to dust grains reflecting the light from the star has been obtained by British astronomers using the Anglo-Australian Telescope in New South Wales—and this is significant, because it is generally believed that the planets in our Solar System built up by accretion from a disk of precisely this type. And despite its faulty mirror, the Hubble Space Telescope has been able to make important contributions, showing that gas-clumps pass in front of the star and cause dramatic changes in the structure of the whole of the cloud.

If our present picture of Beta Pictoris is right, there is a diffuse gaseous disk in a stable orbit round the star, surrounding an inner disk of gas which is drifting slowly inward. There are also streamers of much denser gas, and there is every indication that there are 'clumps' of planetary size.

All of which makes Beta Pictoris very interesting, and as yet it is unique in our experience. We cannot say that it is the centre of a planetary system, much less that any of its planets are inhabited, but it is quite possible. When you look up at the tiny point of light which marks Beta Pictoris, it is not outrageous to think that someone there may be looking up at *us*.

☾7 Scandinavia's New Telescope

La Palma, in the Canary Islands, has become very much of an astronomical centre. It is a Spanish island, but the observatory there is truly international. Atop the extinct volcano of Los Muchachos there are various major telescopes, of which one, the British William Herschel reflector, is the third largest single-mirror telescope in the world. Also there is the Isaac Newton Telescope, transferred from its old site in Sussex; the smaller Jacobus Kapteyn Telescope, the large Swedish solar telescope, and so on. Now there is a new addition, the 100-inch Nordic reflector.

This is a joint effort between Denmark, Norway, Sweden and Finland. Much though I like Scandinavia, I cannot say that it is the best site for astronomical research, and this is why it was decided to come to La Palma, where the conditions are superb. On the other hand, it was clear that a really giant telescope would cost more than the £4,500,000 which was available, and so the organizers opted for a smaller telescope—twentieth in size in the world—with

Dome of the Nordic Telescope at La Palma.

only 25 per cent of the light-gathering power of the William Herschel. To make up for this, important design modifications were introduced. The main essentials were lightness, steadiness and accuracy.

To make this instrument light, the telescope was designed to be very short. Its focal length—that is to say, the distance between the mirror and the point where the images are brought to focus—is only twice the diameter of the mirror itself, so that the telescope looks remarkably squat. This also means that the dome containing the telescope can also be very small, and fits rather snugly around the instrument. The whole design was thoroughly tested to make sure that wind blowing into and round the dome would not cause unacceptable air-currents which would distort the images; all seemed well. Ground heat was avoided by mounting the telescope more than twenty feet above ground level. Of course, the positions and lining-up of the main and secondary mirrors are constantly checked by special computers.

The results are encouraging, and the Nordic telescope is doing all that was expected of it. The Director, Arne Ardeberg of Sweden, has said that he is very satisfied indeed. Researches of all kinds are now in progress, mainly in connection with distant stars and star systems. It just shows what can be done with a relatively limited budget plus a great deal of ingenuity, and I have no doubt that the Nordic telescope has embarked upon a long and fruitful career.

A Look Back to 1833

Some time ago I was lucky enough to buy a copy of *A Treatise on Astronomy*, written by Sir John Herschel in 1833. John Herschel was the son of the discoverer of the planet Uranus, and a notable astronomer in his own right; he visited the Cape of Good Hope during the 1830s, and was really responsible for laying the foundations of southern-hemisphere astronomy (he was also the last man to see Halley's Comet at the return of 1835). Yet some of the ideas in his book seem rather curious now.

For example, consider sunspots. According to John Herschel, they are 'the dark, or at least comparatively dark, solid body of the Sun itself, laid bare to our view by those immense fluctuations in the luminous regions of its atmosphere, to which it appears to be subject . . . . Lalande suggests, that eminences in the nature of mountains are actually laid bare, and project above the luminous ocean . . . . A more probable view has been taken by Sir William Herschel, who considers the luminous strata of the atmosphere to be sustained far above the level of the solid body by a transparent elastic medium, carrying on its upper surface a cloudy stratum which, being strongly illuminated from above, reflects a considerable portion of the light to our eyes, while the solid body, shaded by the clouds, reflects none.'

Sir William Herschel had believed the Sun to be inhabited, which is amazing when you recall that he died as recently as 1822. Sir John did not go as far as that, and he admitted that 'the body of the Sun, however dark it may appear when seen through its spots, *may* nevertheless, be in a state of most intense agitation'. But at that stage it was not definitely known that the Moon is airless, and there was a considerable amount of support for the idea that the corona, seen round the Sun during a total solar eclipse, is due to an atmosphere of the Moon rather than being associated with the Sun itself. This theory was not finally rejected until the middle of the nineteenth century.

John Herschel also held some curious views about the Moon. He believed (correctly) that there were regions on the lunar surface which were very hot when the Sun was overhead, but went on to say that there would be 'constant accretion of hoar frost in the opposite region, and, perhaps, a narrow zone of running water at the borders of the enlightened hemisphere'. Neil Armstrong and the other Apollo astronauts would have been thunderstruck to come across real lunar rivers in which they could paddle!

Saturn's rings, thought Sir John, were probably solid sheets, and any disturbance might cause them to 'precipitate, *unbroken*, on the surface of the planet'. It is also interesting to read his description of Mira Ceti, the variable star which ranges between magnitudes 3 and 10 in a mean period of 331 days. Sir John writes that 'it appears about twelve times in eleven years; remains at its greatest brightness for about a fortnight; . . . decreases during about three months, till it becomes completely invisible, in which state it remains for about five months, when it again becomes visible'. He adds that according to the famous Danzig observer Hevelius, Mira did not appear at all during the four years between October 1672 and December 1676.

Sir William Herschel

This is strange, because Mira never becomes too faint to be seen with a small telescope. I can only conclude that Sir John had never taken the trouble to look for it when it was near minimum brightness. Had he done so, using one of his powerful instruments, he could not possibly have overlooked it.

9 Ring Round the Earth?

Saturn has always been known as the planet with the rings. Until recent years it was regarded as unique, but this is no longer so; all the outer giants have ring systems, though admittedly they are very obscure compared with those of Saturn. Now a new theory by the Danish astronomer Kaare Rasmussen, of the National Museum in Copenhagen, suggests that in the past our Earth also had rings.

What Rasmussen has done is to carry out a survey of all reports of meteor showers and fireballs, together with falls of meteorites, between the years 800 BC and AD 1750 (he stopped at 1750 because after that there were so many observers and so many reports that analyses became impracticable). Between 800 BC and AD 1750 he found that there were well-defined peaks of meteoritic activity, lasting for a few months or a few years, followed by a slow decline before the start of a rise to another peak.

His suggestion is that a peak occurs when either a comet or an asteroid is captured by the Earth and starts to move round us, becoming to all intents and purposes a second satellite. Unlike our natural satellite, the Moon, the

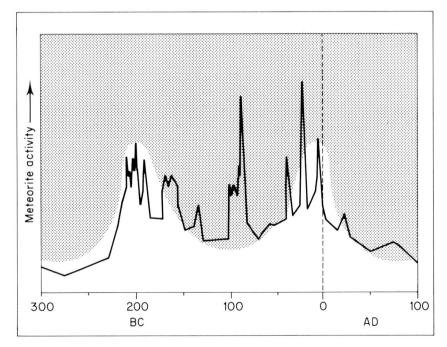

Time-scale of peaks of meteorite activity, according to Rasmussen.

captured body would not last. Instead it would break up into small fragments, and these fragments would be gradually spread around the whole orbit, producing a ring. Some of the particles would come down into the upper air, and we would see an unusual number of meteors; then, as the break-up progressed, more and more particles would come down, and activity would peak. When the supply of particles became exhausted, with the disappearance of the rings, things would revert to normal until the capture of another comet or asteroid, when the sequence of events would begin all over again. Rasmussen puts his 'peaks' at 200 BC, and between 50 BC and year 0, with a secondary peak at 100 BC. He also says that the longest gap was that between 1561 and 1665, so that we might well be due for another cycle.

To me, the main objection to all this seems to be that a wandering comet or asteroid would have to be moving in a very special path to be captured in this way—and even if so, would it break up in the way that Rasmussen assumes? If it were asteroid-sized it would be more likely to remain intact, unless it fell as a mass to produce one cataclysmic event (and remember, it has been suggested that a collision of this sort wiped out the dinosaurs some 65,000,000 years ago). On the whole I admit to being unconvinced, but at least it is fascinating to look at the possibility of an Earth with brilliant, if temporary, rings.

☾10 How Big are the Moon's Seas?

Even with the naked eye it is always worth looking carefully at the Moon, particularly when it is near full phase. The broad dark 'seas' are obvious at a glance. Of course there has never been any water in them, but in the past, they were oceans of lava, so that the name is not entirely inappropriate.

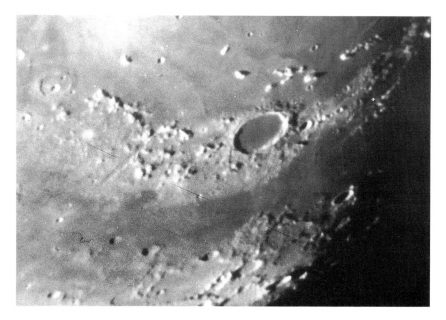

Plato: the dark-floored, 60-mile crater is obvious. The feature to the left is the great Alpine Valley which cuts through the mountain range. Photograph by Commander H. R. Hatfield, 12-in reflector.

The Moon from Galileo. This picture was taken by the Galileo space-craft on 9 December 1990, from a range of 350,000 miles. The Mare Orientale basin (diameter 600 miles) is near the centre; to the upper right is the largest of the lunar 'seas', the Oceanus Procellarum. Reproduced by kind permission of NASA.

Look first at the Oceanus Procellarum or Ocean of Storms, which lies to the left-hand side of the disk of the full moon as seen with the naked eye (assuming that you are observing from the northern hemisphere of the Earth). Its area is 882,000 square miles, which is just about the same as that of Saudi Arabia—and bear in mind that the Moon is a small globe, with a diameter of no more than 2158 miles. The Oceanus Procellarum is not very sharply bordered, but leading off it is the much more definite Mare Imbrium or Sea of Showers, with an area of 333,000 square miles—about the same as that of Pakistan, and slightly larger than Great Britain and France combined.

Mare Imbrium is regular, bordered in the main by high ranges of mountains such as the Apennines and the Alps. The Apennines separate it from the Mare Serenitatis or Sea of Serenity, whose area is 121,000 square miles, equal to that of Italy and larger than Great Britain. It too is regular in outline, and there are not many craters in it; it is one of the smoother lava-plains of the Moon. Of other important seas, the Mare Tranquillitatis has an area equal to that of California, while the Mare Fœcunditatis is comparable with Finland.

It is interesting to look next at the small, well-marked Mare Crisium or Sea of Crises, which lies to the right-hand side of the disk as seen with the naked eye, and is separate from the main system. Its area is 77,000 square miles, equal to that of the American state of Kansas. It seems elongated in a north–south direction, but this is not really the case; the east–west diameter is slightly greater. Because the Mare Crisium lies not far from the Moon's limb as seen from Earth, it is considerably foreshortened.

Foreshortening, in fact, makes the limb regions very hard to map—as I well know, because in the pre-Apollo period I spent many years in attempting it! Because the Moon has an orbit which is elliptical, not perfectly circular,

The lunar Mare Serenitatis, one of the most regular of the 'seas'. Photograph by Commander H. R. Hatfield, 12-in reflector.

it seems to rock very slowly to and fro, so that we can see alternately a little way beyond first one mean limb and then the other. At the extreme limits, a feature may be so foreshortened that it is almost impossible to tell whether we are dealing with a crater or a mountain ridge.

It is also notable that most of the seas form a connected system— Mare Crisium being the exception—and those away from the main pattern are, in general, irregular and patchy. Of special significance is the Mare Orientale or Eastern Sea, right on the limb, and identifiable only under ideal conditions. It extends on to the far side of the Moon, and has proved to be a vast, multi-ringed object. Very little of it can be seen from Earth, and when I discovered and named it, long before the Apollo period, I had no idea how large it was.

Very close to the main limb, beyond the Oceanus Procellarum, lies the dark-floored crater Grimaldi. The interior area is over 11,000 square miles, and if it were more favourably placed on the disk it might well have been classed as a small mare. Is there, then, any basic difference between a 'sea' and a 'crater'? I suspect that there is not.

Finally, look at Plato, in the Alpine region north of the Mare Imbrium. It is 60 miles in diameter, and very regular, with a smooth floor which is so dark that it is instantly recognizable whenever it is in sunlight. It looks elliptical, but in fact it is almost perfectly circular, as photographs taken from space-craft show. Even though Plato lies some way from the mean limb as seen from Earth, the effects of foreshortening are very marked.

C11 Keck: The World's Most Powerful Telescope

Where would you find the world's most powerful telescope? No, not in California, nor in Russia. Today the title has passed to the Keck Telescope, on the summit of Mauna Kea in Hawaii.

Mauna Kea is impressive. It is actually the largest of all terrestrial volcanoes, though much of it lies below sea-level and its summit rises to only about half the height of Everest. Even at this altitude, however, the air is thin, and seeing conditions are generally excellent. Mauna Kea, unlike its neighbour Mauna Loa, is extinct (we hope!) and various great telescopes have been built on it. The Keck, the newest of them, has a mirror 10 metres or 396 inches across—nearly twice the size of the 200-inch Hale reflector at Palomar.

It is virtually impossible to make a single mirror of this diameter—even from a material called Zerodur, which is suitable because it does not expand or contract appreciably as the temperature changes. The Keck mirror is made up of 36 hexagonal segments, arranged together to give the correct optical curve. The telescope, plus the mirrors, weighs 300 tons, but the mirror itself has a weight of only 14.4 tons, as against over 40 tons for the single glass mirror of the huge reflecting telescope in Russia.

Dome of the Keck Telescope on Mauna Kea. When I took this photograph, in 1991, 12 of the 36 segments of the mirror were already in place.

The incomplete Keck mirror, as it was in August 1991.

Making the 36 segments is tricky enough. Fitting them together is even more so, because everything has to be so absolutely accurate. The segments are aligned by a special control system to within a millionth of an inch, which is a thousand times thinner than a human hair.

The basic idea goes back to 1978, and by 1980 it had been agreed to have a segmented mirror instead of a single one, but everything depended upon money (as it always does). The Keck Foundation stepped in with a grant of 70,000,000 dollars, around 80 per cent of the total cost, and work began. The observatory dome was built, and the segments were cast. By the time of my visit a dozen of them were in place, but even this gave a light-grasp greater than that of the Palomar reflector, and the first results were most encouraging. The full telescope should be in working order before the end of 1992.

Whiffletrees for the Keck segments. August 1991, while the mirror was being prepared. A whiffle-tree is a vertical rod which branches out in twelve short posts; together with flex disks, they keep the Keck mirror segments rigidly mounted against sideways motion while still allowing them to be moved for aligning the mirror array. The name comes from a last-century pivoting crosspiece which allowed draft animals in a team to move independently while still pulling a cart.

What will it do? Well, it will tackle almost all branches of astronomy, and it will make use of the most modern electronic devices, which by now have to all intents and purposes ousted the faithful photographic plate. The Keck's enormous light-grasp will enable it to reach out to the furthest depths of the universe, and to study objects so remote that we see them as they used to be thousands of millions of years ago. Another research programme will involve using infra-red detectors to look into the dusty clouds of space which we call nebulæ, where fresh stars are being formed. All in all, there seems no end to what the telescope will be able to achieve.

Even better news came in early 1991, when the Keck Foundation promised another 76,000,000 dollars for a twin telescope to be set up a mere 280 feet from the first. It should be ready by the mid-1990s. The twins will be capable of being used together, and they will have so much resolving power that, in theory, they could distinguish the two headlights of a car, separately, from a distance of 16,000 miles. It's quite a thought.

C12 **Brown Dwarfs**

When is a star not a star? The answer seems to be: 'When it's a Brown Dwarf'. It is too massive to be classed as a planet, but neither can it be regarded as a star, because its core is not hot enough to trigger off the nuclear reactions needed to make it shine in the same way that the Sun does. I have described a Brown Dwarf as a star which has failed its Common Entrance Examination.

Obviously they are going to be dim; they will shine only because they are shrinking, and a certain amount of energy is set free by gravitation. We cannot hope to detect them unless they are fairly close at hand, and there have been a surprising number of false alarms. The first of these came in 1984, when two American astronomers, Donald McCarthy and Frank Low, reported finding a faint companion of the star known as VB8, some 21 light-years from the Earth. The suspected Brown Dwarf became known as VB8 B. (The letters VB stand for Georges van Biesbroeck, a Belgian astronomer who drew up a valuable catalogue of double stars.)

McCarthy and Low assumed that the Brown Dwarf was moving round its parent star, VB8 itself, which is easily visible with a moderate telescope. Because of its dimness, they had to use a new technique called speckle interferometry, which involves taking a series of very short-exposure images and then sorting them out electronically to remove the blurring caused by the Earth's unsteady atmosphere. Everything seemed to be in order. Clearly it was not possible to determine the mass of VB8 B simply by obtaining images of it, but there seemed no reason to suspect any major error.

Before long serious doubts crept in. If VB8 B were cool, as presumably it had to be, it should radiate at infra-red wavelengths. F. Skrutskie, William Forest and Mark Shure looked for it with a very sensitive infra-red detector mounted on the 3-metre telescope on the summit of Mauna Kea in Hawaii, but nothing could be found, and French observers in Chile were equally unsuccessful. VB8 B seemed to have disappeared, and it has never been found again.

Speckle interferometry is a delicate technique, and could produce misleading images; alternatively, what was seen was merely a cloud of gas or small particles. There is also a slim chance that VB8 B is larger and faster-moving than was thought, and has now lined up with VB8 itself, in which case it will eventually reappear. Meantime, it must be listed as 'missing'.

In 1988 a more promising candidate was found by infra-red techniques, again from Mauna Kea but this time by Benjamin Zuckerman and Eric Becklin. It is associated with a very old White Dwarf, Giclas 29-38, which is 46 light-years away. The Brown Dwarf has an estimated diameter of 125,000 miles, which is larger than the White Dwarf around which it revolves. The Brown Dwarf seems to be fairly dense—several tons of its material could be packed into an eggcup—but not nearly as dense as the White Dwarf, which is so highly evolved that it has exhausted all its nuclear 'fuel' and has collapsed. Of course, the White Dwarf is much the more massive of the two, and indeed its mass is equal to that of the Sun.

All this is very interesting, but it is not conclusive, and the same can be said of results based upon what is termed the astrometric principle. A relatively lightweight star may be affected by the pull of a companion, so that it seems to 'wobble' very slightly as it moves along; the amount of the 'wobble' gives a key to the mass of the companion. In America, David Latham and his team have found effects of this sort with a dim star, HD 114762, which is 90 light-years away and is not too unlike the Sun, though distinctly older. Latham believes that the companion is about ten times as massive as Jupiter, the senior member of the Sun's family, and is close to its parent star—about as close as Mercury is to our Sun. Rather different investigations by Bruce Campbell and his colleagues

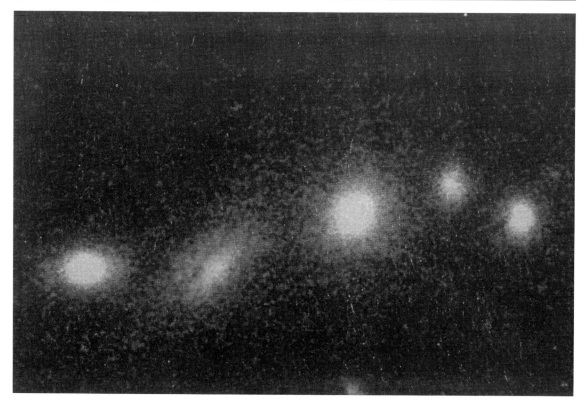

Chain of five galaxies

in Canada indicate that several stars may have companions of between 1 and 10 times the mass of Jupiter, in which case they must qualify as Brown Dwarfs.

The next and, so far, latest claim has been made by Mike Hawkins at Edinburgh, who has been examining plates taken with the Schmidt telescope at Siding Spring in Australia. One star was of special interest to him. It was faint, reddish, and over a period of two years, from 1988 to 1990, it seemed to have shifted against its background. From this it has been calculated that it must be about 68 light-years away, and to have a luminosity of only 1/20,000 that of the Sun, with no more than 5 per cent of the solar mass. If this is correct it is some 30 per cent dimmer than the previously faintest-known star, and could be one of our 'missing links'. To quote Dr Hawkins:

'From the colour of this star, it was believed to be very red, very cool and probably very faint. Our measurements indicate that it is even fainter than imagined. We think that it is the faintest brown dwarf candidate to date.'

It must be a curious place. I would hate to try to walk on its surface (it would be too warm for that), but it will be totally unlike either the Earth or the Sun. If it really is a Brown Dwarf, there may be many others which we cannot see simply because they are too dim.

To become a true star the mass must be at least 8 per cent that of the Sun. Hawkins' star is well below this limit, so that it will never become hot enough at its core to shine by nuclear reactions. Perhaps, at last, we have really identified our first Brown Dwarf.

☾13 A Near Miss in Space

Many thousands of people per year visit Cape Canaveral, the American rocket base. It is not nearly so easy to travel to Baikonur, the Russian equivalent, which is in a desolate part of what used to be the Soviet Union—in Kazakhstan, in the general area of the Sea of Aral and not far from the border with Afghanistan. Until 'glasnost' its very identity was more or less secret, and it did not appear on maps. It is nowhere near the village of Baikonur; the railway station is Tyuratam, and the airfield is Krainj. The large city associated with it is Leninsk, which is completely new. It has a population of 100,000, every one of whom is connected with the space programme in one way or another. There are no roads to Leninsk, so that everything has to be brought in by rail or air—even food. I went there from Moscow on a military plane, which was not exactly luxurious (wooden seats, no toilets!) but at least it landed on time.

(Incidentally, will Leninsk be renamed now that the USSR has disintegrated? I would not be surprised. The great Russian rocket genius was Sergei Korolyev, so perhaps 'Korolyevville' would be a good replacement.)

Baikonur is totally unlike Cape Canaveral. Even in the spring, when I was there, it is icy and cold; it is also very spread-out, so that driving from one complex to another takes considerable time. I was able to go to all the important launch-pads, including that named in honour of Yuri Gagarin, the first of all spacemen. There was the vast Energia complex, controlling what is now the world's most powerful rocket; it is the launcher of Russia's equivalent of the Shuttle, and has been used to launch Buran, which means 'Snowstorm'. Up to that time Buran had had two flights, the first of which was purely

The Energia launch-pad at Baikonur, 1991.

Cape Canaveral. A Shuttle in its launch-pad; a warm day, with green grass all around!

automatic and both of which had been successful. Many lessons had been learned from the various American problems with their Shuttle vehicles.

While I was there, I was able to see an actual launch. The space-station Mir was in orbit round the Earth, with two cosmonauts on board, Viktor Afanasyev and Musa Manorov. Supplies are taken up by unmanned rockets of the Progress series, and it was one of these, Progress-7, which I saw starting on its journey. It was most impressive; there was a brilliant glow, and then a thunderous roar as the rocket climbed into the sky, soon to be lost to view. Launches of this kind are now routine, though it was interesting to find that the audience included a number of richly bemedalled Soviet generals.

Yet there was a hitch, about which I knew nothing until I was back in England. The Progress vehicle was scheduled to dock with the space-station, but the first attempt failed, and there had to be a delay before the supply ship and Mir were suitably placed for another try. As the Russians admitted during their television programme on 28 March, it was touch and go. Only just in time, the ground controllers realized that the orbit of Progress was not going to bring it in to a gentle docking with Mir; it was on a direct collision course. When the two space-craft were a mere 60 feet apart, the ground controller had to override the computers and alter the path of Progress-7. The on-board motors fired, and Progress by-passed Mir at no more than forty feet. If there had been a collision, the cosmonauts would certainly have been killed. As the Soviet commentator said, 'Imagine a heavy truck heading at full speed for a gate which is not real, but is only painted on wall!'

The story had a happy ending. Afanasyev and Manorov were to carry out repairs which allowed the cargo module, with its vital supplies, to dock safely, so that the whole mission went on as planned. But it was a very close shave indeed, and yet another reminder that space-travel is a risky business.

Launch of a Progress unmanned rocket to Mir. I watched this launch, which was successful—the problems came later.

After leaving Baikonur I went to Star City, where the cosmonauts were trained, and where there is a full-scale mock-up of Mir. To be candid, Star City is not imposing (there is nothing of Professor Quatermass about it), and this is typical of the whole Soviet approach. In Western eyes, some of their equipment—particularly in the field of computers—is primitive, but it works, and this is what matters.

Since my visit, of course, there has been the abortive coup aimed at toppling President Gorbachev. By now the old USSR has more or less fragmented, and what will happen to the space programme is anybody's guess. I very much hope that it will continue, because, surely, exploring space is one of the very best ways of uniting the nations of the Earth; but we must wait and see.

 **Some Oddities**

Every science produces its quota of odd episodes, and astronomy is no exception. For example, investigations can be brought to a grinding halt in quite unexpected ways. This happened in the early 1980s, when a team of researchers at the Ohio State University in America used a radio telescope to search for intelligent signals from outer space. This was a perfectly logical and serious project, and it involved using a very powerful computer. In this case the computer available was old, but still extremely efficient. Unfortunately, a mouse elected to build a nest at the intake of the computer's disk drive, thereby cutting off the air supply. This made the disk drive destroy itself, and the computer was too old to be fitted with a replacement, so that the entire programme had to be abandoned.

Mice are bad enough, but did you know that a major telescope was once put out of action by a fly? It happened in 1984 to the 26-inch refractor at the Royal Greenwich Observatory. The fly made its way into the telescope tube, and died. It then fell on to the cross-hairs of the sighting eyepiece, and broke them. Disaster! How was it possible to extract the fly and then repair the cross-hairs, which are usually spider's webs? (One brave Australian used to make the threads from the webs of black widows.) Conferences were held. There was a carefully planned fly-removing operation, and then Norman King, an electronics craftsman, managed to replace the broken cross-hairs with nylon filaments, which are only 20 microns in diameter. One micron is equal to a millionth of a metre; a human hair, 70 microns in diameter, is much too coarse to satisfy an astronomer.

A different sort of oddity dates back to 1962. Scene: the Haute Provence Observatory in France, where two eminent astronomers, D. Barbier and N. Morguleff, were busy studying the spectra of some dwarf stars. Suddenly they detected bright spectral lines which indicated flares due to the element potassium. Now, potassium flares are the last things to be expected in stars of this sort, and the French paper on the subject caused a tremendous amount of interest, particularly when the observations were repeated. Yet other searches for potassium flares of the same type were negative, and astronomers in general were frankly puzzled.

The mystery was solved in 1967 by astronomers at the University of California. What had happened was that either during or (more probably) after the photographic exposure, one of the astronomers had struck a match in order to light his cigarette. The matchlight had been able to find its way into the equipment—and matches, of course, contain potassium . . . .

Next, the Ogre. One night in August 1984 a trio of Canadian astronomers— Bruce Walters, Kai Millyard and Bill Katz—carried out a meteor watch. Suddenly they detected strange flashes of light coming from the direction of the constellation of Aries, the Ram, and other similar observations followed. What could be the cause? One theory involved the very short-wavelength radiations known as gamma-rays, and it was suggested that the flashes might come from an *O*ptical *G*amma-*R*ay *E*mitter (OGRE). The affair became headline news in astronomical circles, and the photographs, published in various journals, looked very interesting indeed.

Then, alas, the cause was found. It was simply a spinning artificial satellite, which happened to be moving in a way which made it cross the Aries region at regular intervals. Its irregularly shaped body reflected the sunlight, so that it flashed; that was all. The Ogre disappeared from the list of astronomical objects to be studied, and the prestigious journals did their best to forget all about it.

Next, a real curiosity. In August 1987 Dr José Arguelles, an art historian, found that all the planets would line up in a way which happens only once in every 23,412 years (please do not ask me how he arrived at this figure). Using astrology, Mayan and Aztec lore, he concluded that it was essential for 144,000 people to 'resonate together'—that is to say, hum—in order to avert disasters which would involve a galactic beam and a deluge of flying saucers. People listened, and thousands of Americans went to the tops of mountains and hummed. The flying saucers stayed away and the world is still here, so that presumably the manœuvre was successful. I assure you that this story is actually true!

Few objects in modern times have caused so much excitement to astronomers as the supernova which flared up in the Large Cloud of Magellan in 1987. It was the only naked-eye supernova since 1604, and at a distance of a mere 169,000 light-years it was close enough to be studied in some detail.

It was a Type II supernova, involving the collapse of a very massive star (and in this case the progenitor star was known; it was a blue supergiant, not a red one, which was the first surprise). After such an outburst, the core of the old star shrinks into a very small, super-dense body made up of neutrons, spinning round quickly and sending out the pulsed radio waves which have led to the name of 'pulsar'. Everyone expected that a pulsar would form from the wreck of the 1987 supernova. And on 18 January 1989 astronomers at the Cerro Tololo Observatory, in Chile, found it—so they thought.

Using the great 4-metre reflector, a team headed by John Middleditch recorded a flashing object in just the right position. It looked like a neutron star candidate. It was under observation for seven hours that night, and one astronomer commented that 'We were waiting for the egg to hatch, but when it did, instead of a chicken we got a Ferrari turning over at 120,000 r.p.m.!'.

The amazing fact was that the pulsar seemed to be spinning at a rate of 1968.63 turns per second, much faster than any other known pulsar. It was also surprisingly bright, at a peak of magnitude 18 (the magnitude of the gaseous remnant at that time was about 12).

But the observations were not repeated, and the situation began to look very peculiar. The pulsar was unlike anything known to science, and during the next few months all sorts of theories were circulated.

One suggestion was that there was a companion body orbiting at about a million kilometres from the pulsar, well inside where the progenitor star's surface used to be. If the flashes came from the actual pulsar, the maximum diameter of the body could be no more than 25 kilometres, as otherwise the equatorial regions would be spinning round at a speed faster than that of light—which is impossible. Norman Glendinning, of the Lawrence Berkeley Laboratory, then proposed that the pulsar had 'turned round' so that its beams of radiation no longer passed, lighthouse-like, across the Earth. Then Joshua Frieman and Angela Olinto of Illinois proposed that the object was made up of 'strange matter', strong enough to withstand forces which would pull an ordinary neutron star apart.

The Irénée du Pont reflector at Las Campanas.

Supernova 1987a, in the Large Cloud of Magellan, as seen from the ESO's Faint Object Camera on the Hubble Space Telescope. The supernova is shown in the upper left-hand frame, and an unresolved comparison star from the same exposure to the upper right. The lower frame shows the luminescent ring around the supernova. Reproduced by kind permission of ESA and NASA.

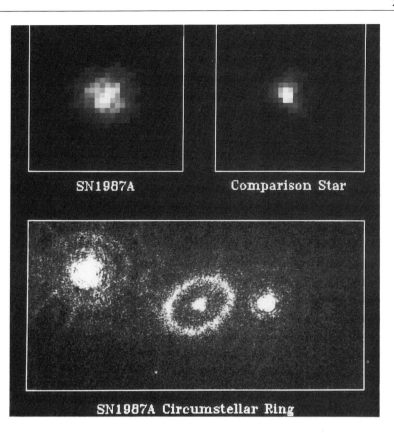

'Light echoes' from the supernova 1987a in the Large Cloud of Magellan. A double light echo from the supernova was observed on 13 February 1988 with the ESO's 3.6-m telescope and the EFOSC instrument by Dr Michael Rosa. The light echoes are reflections in interstellar clouds in the Large Cloud of Magellan from the supernova outburst of 23 February 1987. The light echoes are seen as two concentric rings round the over-exposed image of the supernova itself. Reproduced by kind permission of the European Southern Observatory.

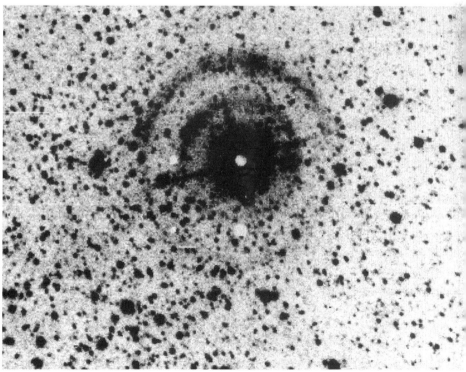

Alas and alack! Further investigation showed that the flashes were due not to anything in the Large Cloud of Magellan, but to part of the mechanism of the Cerro Tololo telescope. It was all somewhat embarrassing, but the astronomers concerned promptly and generously admitted their mistake. Up to the present time no pulsar has been found in the supernova remnant.

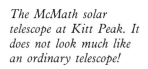

☾15 **Sunquakes**

The McMath solar telescope at Kitt Peak. It does not look much like an ordinary telescope!

If you have ever observed the Sun by means of a telescope, you will have noticed that the dark patches which we call sunspots move slowly across the bright disk, showing that the Sun is rotating on its axis. It takes a spot nearly a fortnight to cross from one edge of the Sun to the other, so that the spin is much slower than that of the Earth; it takes the Sun 25.3 days to complete one full turn if we reckon from the solar equator, and at least a week longer if we reckon from the poles, so that the Sun does not rotate in the way that a solid body would do. Remember, its surface is gaseous, with a temperature not far short of 6000 degrees Centigrade. Sunspots are not permanent; small ones may have lifetimes of only a few hours, while even large groups seldom persist for more than a few months even though this is long enough for them to make several crossings of the Earth-turned side of the globe.

Deep inside the Sun, near the core, energy is being produced by the conversion of one element (hydrogen) into another (helium). It has been widely believed that the core must spin more quickly than the surface, but this view has now been challenged by two astronomers, John Harvey and Tom Duvall, working at the Kitt Peak Observatory in Arizona, where there is the world's most powerful solar telescope—which does not look like a telescope at all, but gives the impression of being an inclined tunnel; the Sun's rays are caught by the mirror at the top end, and are reflected down the tube, eventually arriving in the solar laboratory.

Harvey and Duvall have played a great part in founding the new science of helioseismology—that is to say, the study of sunquakes. Forces inside the Sun make the surface tremble, like the sides of a metal bell which has been struck: the solar atmosphere pulsates at the rate of hundreds of feet per second. Harvey and Duvall have studied two dark lines in the orange part of the Sun's spectrum, and have found that as the surface ripples up and down these lines shift in wavelength, so that the extent and the frequency of the vibrations can be found. These oscillations are really made up of several different 'modes', each one referring to a different depth below the Sun's surface. Therefore, the Kitt Peak astronomers have in effect been able to 'peel back' the Sun's layers and see what is happening below the visible surface.

The results are unexpected. The rate of rotation speeds up just below the surface, but then slows down again in what we call the convective zone, where sunspots have their roots. Most of the Sun spins at this rate; the core may

Projecting the Sun. Point the telescope toward the Sun, without looking through it, and then project the image. In this picture a projection box is fixed to the telescope, but it is quite adequate simply to hold a screen behind the eyepiece.

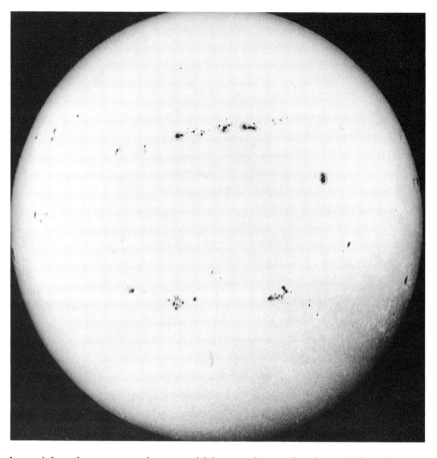

Sunspots: an active solar disk.

be quicker, but not nearly so rapid in rotation as has been believed up to now. This could mean more 'mixing' of the material in the Sun, which will, so to speak, refuel the energy-producing core. The upshot is that the life expectancy of the Sun in its present form may be rather longer than most people have expected.

This is all very speculative and uncertain, but it could turn out to be very important. Moreover, the Sun is a normal star, and in learning more about it we are also learning more about the other stars of the Galaxy.

Watching sunspots is an enjoyable pastime, but one thing must always be borne in mind: never, under any circumstances, use a telescope or binoculars to look direct, even with the addition of a dark filter. Focusing the Sun's light and heat on your eye is certain to do serious damage, and will probably result in blindness. This is not mere alarmism: it has happened. The only sensible method to look at sunspots is to use the telescope as a projector, and send the Sun's image on to a screen fixed or held behind the eyepiece. Most regrettably, some manufacturers of small telescopes sell what they call sun-caps, to be used for direct viewing. If you have one of these sun-caps, I recommend that you take it into the garden, put it on the ground, and hit it very hard with a sledge-hammer until it has been reduced to fine, harmless powder.

C16 Goodbye, Solar Max

We often regard our Sun as a steady, well-behaved star, but in fact it is to some extent variable. Every eleven years or so it is at the peak of its activity, with many sunspots and spot-groups; there are violent solar flares (unfortunately, not usually observable without special equipment), and brilliant displays of auroræ or polar lights, which are due to electrified particles sent out from the Sun which crash into the Earth's outer layer and cause the lovely glows which we see.

The Sun was near the peak of its cycle at the start of the 1980s, and again at the start of the 1990s. (The period is not absolutely regular, and neither are all maxima equal in intensity.) At spot-minimum, the disk may be blank for many days consecutively. It cannot honestly be said that we yet have a perfect understanding of how the Sun 'works', but space research methods are giving us a great deal of new information, and on 14 February 1980 a special satellite was launched from Cape Canaveral in Florida. It was officially called the Solar Maximum Mission (SMM), but everyone referred to it affectionately as Solar Max. It was designed to stay in orbit for at least ten years, to send back data about the Sun through one complete cycle.

Solar Max: artist's impression.

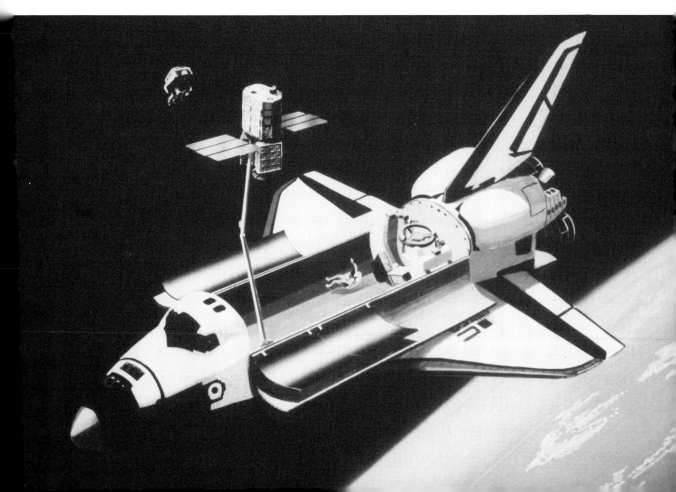

It began well, but then there were electrical problems, and Solar Max became unreliable. This led to an amazing 'rescue'. Astronauts in the Space Shuttle went aloft, fished Solar Max into the repair bay, put it right, and re-launched. This was the first time that anything of the sort had been attempted, and the whole operation went smoothly from start to finish.

Solar Max was back in business, but it was not in a permanent orbit, because it was not quite above the last traces of the Earth's air. This meant that it was affected by friction. Moreover, during solar maximum the upper atmosphere thickens slightly, and this in turn increases the problems due to friction. It was realized that unless something was done, the satellite would spiral downward, enter the denser air, and break up, possibly showering fragments over a wide area—as the Skylab space-station had done years before.

Another rescue mission could have been carried out, and solar scientists pleaded for it. There was nothing wrong now with Solar Max itself, and it was providing streams of valuable data; if it could be given an extra 'boost' and sent into a higher orbit, it could continue working for years more. Alas, this was a time of financial cut-backs, and those who controlled the funds were not to be convinced. No extra grant was forthcoming, and so, to the universal distress of astronomers all over the world, Solar Max was allowed to die. On 2 December 1989, almost ten years after having been launched, it burnt up over the Indian Ocean, at 3.1 degrees north, 88.6 degrees east. Luckily the fragments came down over the sea, and there was no damage to life or property.

It was a sad ending, and surely it should never have happened. The satellite could have been saved at a cost less than that of one of the Patriot missiles fired during the subsequent Gulf War. True, Solar Max had a useful life of almost a decade, but it ought still to be in orbit.

☾17   Danger from Chiron

In 1977 Charles Kowal, using the large Schmidt telescope at Palomar in California, discovered a most unusual asteroid, subsequently numbered 2060 and named Chiron in honour of the wise centaur of the Golden Fleece legend. Most asteroids keep strictly to that part of the Solar System between the orbits of Mars and Jupiter, but Chiron proved to be much further out, and to spend almost the whole of its 50-year revolution period between the orbits of Saturn and Uranus. This is the last place where one would expect to find an asteroid, and in fact Kowal was not looking for one when he found it. Moreover Chiron is larger than most members of the main swarm; its diameter is at least 150 miles, and probably well over 200. It will next reach perihelion, its closest point to the Sun, in 1995, and it will then be of the 15th magnitude, so that it will be within the range of good amateur telescopes. The distance from the Sun ranges between 794,000,000 miles and 1,757,000,000 miles. Calculations show that in the year 1664 BC it approached Saturn to within

Chiron, photographed with the 48-in Schmidt telescope at Palomar soon after its discovery in 1977. The stars appear as small disks, but Chiron moved appreciably during the exposure, so that it left its trail on the plate.

ten million miles, which is not much greater than the distance between Saturn and its outermost satellite, Phœbe.

Chiron was a puzzle from the outset. Then, in 1988, it caused a major surprise when it started to brighten up, appreciably if not spectacularly. Next, in 1990, photographs taken from the Kitt Peak Observatory in Arizona showed that it appeared 'fuzzy'; in fact it had developed a comet-like coma, possibly made up of a mixture of carbon dioxide and dust which had been frozen on the surface and was now warming up as Chiron drew inward toward the Sun.

Could Chiron be a huge comet rather than an asteroid? It was a possibility, but the size was too great—and remember that even the nucleus of a major comet such as Halley's is less than 20 miles across. All in all it seemed more likely that Chiron was a true asteroid, or even a planetesimal (a surviving remnant of the building blocks which condensed to form the outer planets) and that the 'fuzz' represented nothing more significant than a temporary atmosphere. Now, however, the Manchester astronomers Mark Bailey and Gerhard Hahn have proposed that Chiron is not only cometary, but is a vast unguided missile which could one day harm us.

It was already known that Chiron's orbit is unstable. Bailey and Hahn believe that the object was once much closer to the Sun than it is now, and was much warmer, so that parts of its icy body were broken off and moved away from the main mass to produce a stream of meteors—the Taurid stream—through which we pass every June and November. It is also suggested that it was a fragment of Chiron which hit the Earth 65,000,000 years ago, sending up so much dust and causing such a change in our climate that the dinosaurs, which had ruled the world for so long, could not cope with the new conditions, and died out. If so, then Chiron could again swing inward and produce another catastrophe of the same kind.

I am always dubious about impact theories involving dinosaurs, and I am not in the least apprehensive about Chiron. Even if it does swing inward at some future time, it cannot be within range of us for many, many thousands of years. It remains an enigma, but I agree with Bailey and Hahn when they say: 'It is such an important object that if one is going to learn anything about it, it would be marvellous to land on it.' Certainly a space mission to Chiron would be spectacular!

18 Tinkering with Time

Not many people today remember the name of William Willett, who lived from 1856 to 1915 and spent much of his career at Chislehurst in Kent, but he has had a great effect on the lives of people who live in Britain, because it was he who introduced the idea of Summer Time. Apparently the notion came to him one morning, as he was riding on horseback over St Paul's Cray Common before beginning work.

Willett realized that putting clocks forward by one hour during summer months would provide lighter evenings at the expense of darker mornings, and this principle of 'daylight saving' was finally adopted, though farmers in particular objected to it. During the war we even had Double Summer Time, and I well remember that once, when I came home on leave from my flying duties in the RAF, I happily played outdoor tennis at midnight. The experiment of keeping Summer Time throughout the year was also tried during the 1980s, and it may well be restored in the future.

The fact that the Earth goes round the Sun not in 365 days, but in 365¼, makes our calendar somewhat complicated. A few unusual ideas have been proposed from time to time. There was, for example, the Calendar of the French Revolution, which was in use in France from 1792 to 1806; there were twelve months of 30 days each, and the 'hour' was about twice as long as the conventional hour. Much more recently, in 1989, the Leeds City Council, which is fairly typical of the Loony Left, introduced what they called 'Metric Time' for the benefit of their employees. The result defied description!

Yet the length of the day really is increasing, and we now have clocks which are better timekeepers than the Earth itself. As everyone knows, the tides are due mainly to the pull of the Moon, which leads to friction between the water and the ocean bed, with the result that the Earth's rate of spin is slowing down; also, the Moon is slowly receding from the Earth. The average increase in the length of the day is 0.00000002 of a second, but over a sufficiently long period it adds up, and becomes evident in the timing of eclipses which happened long ago. A very simple calculation will make this clear.

As each day is 0.00000002 of a second longer than the previous day, then a century (36525 days) ago the length of the day was shorter by 0.00073 second.

Taking an average between then and now, the length of the day was half of this value, or 0.00036 second, shorter than at present. But since 36525 days have passed by, the total error is $36525 \times 0.00036 = 13$ seconds. This means that the position of the Moon, when 'calculated back', will be in error; it will seem to have moved too far, i.e. too fast. Unless we take this 'secular acceleration' into account, our calculations of past eclipses will be wrong—and this is precisely what we find.

Going back still further, we can deduce that in the remote past the days were very much shorter than they are now. Corals give us valuable clues, because they develop 'rings' as they grow, producing one ring per day. In the 1960s John Wells, of Cornell University in America, found that fossil corals dating back around 370,000,000 years had produced 400 daily growth rings per year. The obvious inference was that there were then 400 days in the year, and each day was a mere 21.9 hours long. Mind you, there were no astronomers to check on what was happening; this was during the geological period known as the Devonian, and the most advanced life-forms on Earth were amphibians, who were not in the least interested in science.

Even earlier—around 800,000,000 years ago—we depend upon stromatolites. These are domes and columns of limestones, produced by the remains of tiny living things. Stromatolites are very ancient, though some are still to be found today, notably in Western Australia. The ways in which they grow and develop makes them useful for dating purposes, and it seems that 800,000,000 years ago there were 435 days in every year, so that the length of the day was no more than 20.1 hours. At that epoch, however, there was no life on Earth except for very lowly forms in the seas.

It would be foolish to claim that we can be precise about all this, but the basic principles are definite enough. And the days are still getting longer.

C19 The Case of the Killer Tomatoes

While on a recent visit to the United States, I came back late to my hotel after a meeting and decided to watch television. I was privileged to see *Plan Nine from Outer Space*, which is universally acclaimed as being not only the worst film ever made in Hollywood, but incomparably the worst film ever made anywhere. From all accounts its nearest rival is *Attack of the Killer Tomatoes*, dating from 1978. Unfortunately I have never seen it, but I was reminded of it by a curious episode which took place in 1990.

Six years earlier, an artificial satellite had been launched which carried tomato seeds. This was a NASA experiment to see whether seeds exposed to cosmic radiation would produce plants with unusual characteristics. There were about twelve and a half million seeds altogether. Originally they were meant to be left in orbit for only ten months, but various delays meant that it was only in 1990 that the crew of the space-shuttle *Columbia* fished them down, so

Splashdown of the Apollo 12 astronauts. They were promptly quarantined— but this was the last time that quarantining was regarded as necessary for men returning from the Moon. Reproduced by kind permission of NASA.

that they had been in space for six years and had absorbed much more cosmic radiation than had been intended.

Then a memorandum from a NASA contractor was leaked to the Press, saying that there was 'a remote possibility that radiation-caused mutations could cause the plants to produce toxic fruit'. Killer tomatoes, indeed! But schools all over America had already been sent the seeds, and over four million pupils were taking part in the experiment, while there were also some seeds in English schools, and a few had been planted at Television Centre in London.

There was general consternation. A Miss Rachel Mable, of the University of California, said 'I think it's a great project, but I can't recommend it to any more schools. Kids will eat anything that will grow.' An official of the Park Seed Company, which had provided NASA with the seeds, 'seemed to favour against' eating the tomatoes. On the other hand Robert Brown, Director of NASA's external affairs division, said 'I don't think we've got a killer tomato plant here. That's Hollywood script stuff.'

Tracking down all the twelve and a half million seeds appeared to be a rather difficult task, but at Selston, in Nottinghamshire, some were certainly in the hands of children of the Holly Hill Primary School, and the Ministry of Agriculture took firm action. Every tomato, they said, must be burned. The soil must also be destroyed, and even the pots in which the tomatoes had been grown.

Probably we can see here the influence of the first really popular television space fiction film, *The Quatermass Experiment*, in which two space-travellers return in the form of a merged hybrid creature with murderous tendencies. All Britain watched it (I did!) and of course it is true that far-fetched though this sort of thing may be, we do not yet have any positive knowledge of the long-term effects of being exposed to cosmical radiation. There are precedents, too; the first Moon travellers were quarantined on their return to make sure that they had not been contaminated, and quarantining was abandoned only when it became patently clear that the Moon is, and always has been, absolutely sterile. The situation may admittedly be less clear-cut with Mars, which does have an atmosphere. Meanwhile, the edict has gone forth. The Ministry granted Holly Hill a three-month licence to grow their tomatoes; this period has now expired—and so the tomatoes, soil and plants must go. Yet another scientific experiment has come to a premature end!

C20 Celestial Crosses

Almost everyone must have heard about the constellation of the Southern Cross, even though it is never visible from Europe or the main part of the United States; it is as familiar to Australians and South Africans as the Great Bear is to Britons. Yet it is not the only cross in the sky. There are two more, and it may be of passing interest to compare them.

Crux Australis, the Southern Cross itself (now known simply as Crux) is actually the smallest constellation in the whole of the sky; there are 88 accepted constellations altogether, and in its dimensions the Cross comes 88th. (The No. 1 position used to be occupied by Argo Navis, the Ship Argo, but it was so unwieldy that finally the International Astronomical Union lost patience with it, and chopped it up into a keel, sails and a poop.) Crux

Crux Australis—the Southern Cross—as I photographed it with a hand held camera from South Africa.

was not even recognized as a separate constellation until 1679, when it was formed by an astronomer named Royer. Before that it had been included in Centaurus, the Centaur, which surrounds it on three sides. The brilliant leaders of Centaurus, Alpha and Beta, are known as the Pointers to the Cross.

The four leading stars of Crux are Alpha or Acrux (magnitude 1), Beta (1½), Gamma (just below 1½) and Delta (just above 3). These four are arranged in a shape which resembles a box or a kite. There is a fifth star, Epsilon Crucis (magnitude 3½) roughly between Alpha and Delta. It does not really spoil the main pattern, but it is nowhere near the middle of the 'box'. There is no central star at all—and impressive though it may be, Crux is not in the least like a cross.

Two of the four main members are notable. Alpha Crucis is a lovely double, separable with a small telescope, with a third star in the same field. Gamma Crucis, third in order of brilliancy, is a red giant, whereas the other leading stars are hot and bluish-white. This is obvious even with the naked eye, and binoculars bring out the contrast splendidly.

Many unwary observers have been deceived by the False Cross, which lies approximately between Crux to one side and the brilliant star Canopus to the other (though admittedly it is some way off the line joining the two). The False Cross lies partly in Carina, the Keel of the old Ship, and Vela, the Argo's sails. Its shape is very like that of Crux. The four principal stars are Epsilon Carinæ (Avior), Iota Carinæ (Tureis), Kappa Velorum (Markeb) and Delta Velorum (which has no accepted Arabic name; in any case, proper names of stars are hardly ever used nowadays, except for stars of the first magnitude and a few special cases such as Polaris and Mira). All four are between magnitudes 1.8 and 2.5, so that the overall aspect is more symmetrical than that of Crux, and the whole group is larger as well as being less brilliant. As with Crux, one of the main stars is red—in this case Epsilon Carinæ— while the other three are white.

Both Crux and the False Cross are immersed in the Milky Way, and the same is true of our third candidate; Cygnus, the Swan. This is a northern constellation, and indeed part of it is circumpolar from Britain—that is to say, it never sets.

This time we really do have an X pattern. The leading star, Deneb, is of the first magnitude, and is a true cosmic searchlight at least 70,000 times as luminous as the Sun. The centre of the X is marked by Sadr or Gamma Cygni (magnitude 2); to the left and right are Delta and Epsilon Cygni. The remaining member of the pattern, Albireo or Beta Cygni, is fainter and further away from the centre than the rest, and so upsets the symmetry, but to make up for this Albireo is a lovely double, with a golden-yellow primary and a blue companion. The separation is 35 seconds of arc, so that both members of the pair may be seen with a small telescope or even powerful binoculars.

Cygnus is often nicknamed the Northern Cross, for reasons which are obvious as soon as you look at it. Deneb makes up a large triangle with Vega in Lyra and Altair in Aquila; many years ago, during a television broadcast, I introduced the nickname of 'the Summer Triangle' for these three, and everyone now seems to use the term, even though it is completely unofficial and the three stars of the Triangle are not even in the same constellations.

C21 The First Map of Venus

Before the Space Age we knew almost nothing about the surface of Venus. We can never see it, because it is always hidden by the dense, cloudy atmosphere; we did not even know the length of its rotation period, which was generally assumed to be of the order of a month (actually it is 243 days, longer than Venus' 'year' of almost 225 Earth days). It is therefore rather surprising to find that the first map of Venus was drawn as long ago as 1726!

Its author was Francesco Bianchini, who lived from 1662 to 1729. He was born at Verona, becoming librarian to Pope Alexander VIII, and was an energetic observer, mainly from Rome. He made recognizable drawings of lunar formations, but is now remembered chiefly for his book *Hesperi et Phosphori Nova Phænomena*, completed shortly before his death. Using one of the small-aperture, long-focus refractors of the day—it had a 2½ inch object glass, and a focal length of 150 feet—he drew up a chart of the markings which he had observed on Venus, and even named them.

Naturally enough, he regarded the dark areas as seas and the bright regions as land. Altogether he recorded seven seas, together with straits and promontories, and was quite confident that his chart was accurate. The names are fascinating, for example his 'first sea' was named the Royal Sea of King John V of Algarvia and Lusitania, while the second was named after the Infante Henry, fifth son of King John I—better remembered today as Henry the Navigator, who never ventured far himself, but dispatched expeditions to all known parts of the world, and founded colleges of mathematics and science. The third sea was named after King Emmanuel; the fourth after Prince Constantine; the fifth after Christopher Columbus; the sixth after Amerigo Vespucci, who gave his name to America, and the seventh after Galileo, for

Bianchini's Telescope: old engraving.

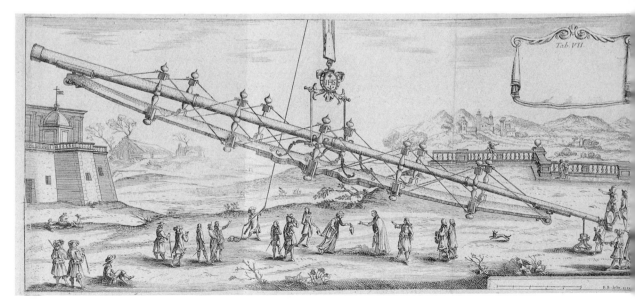

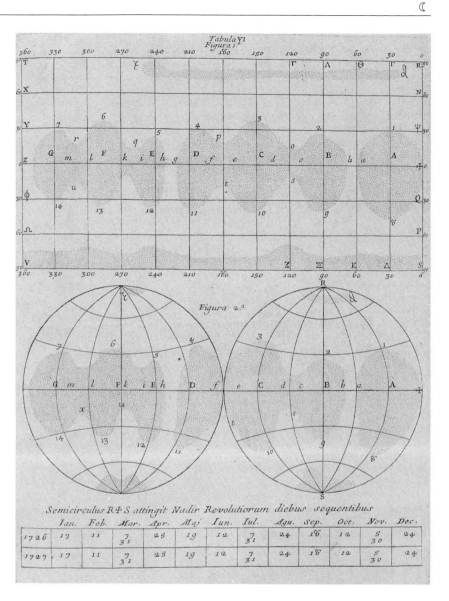

Bianchini's map of Venus.

whom Bianchini had the greatest admiration. Straits were named after Albuquerque (Alphonso de Albuquerque, who undertook the first Indian expedition under King Emmanuel in 1503); Francisco de Almeida; Nunni da Cunha; Vasco da Gama; Duartes Pacheco; John de Castro; Ferdinand Cortez, and Francisco Pizarro. Giovanni Cassini, undoubtedly the greatest of the early planetary observers, was also allotted a strait.

I am sure that you will recognize some of these names—and equally sure that others will be unfamiliar. In our own age it is not always easy to remember that Portugal and Spain were once the leaders of world exploration.

What did Bianchini actually see? That he was honest and industrious is not in question, but his unwieldy telescope with its poor light-grasp would

On Venus: a Russian drawing of a Venera probe standing on the planet. A far cry from Bianchini's view of it!

have been inadequate to show markings on Venus even if they had existed. The problem is, and always has been, that the brilliance of Venus, plus the fact that when 'full' it is out of view on the far side of the Sun, makes visual work very difficult even with modern instruments. Bianchini's 'seas', 'straits' and 'promontories' were no more than optical effects.

Today we have reliable maps of Venus, thanks to the radar equipment carried on space-craft such as Magellan. There are no seas; instead we have craters, valleys, uplands and (probably) active volcanoes. Bianchini had no means of knowing this—and yet we must respect him as a pioneer observer, who did his best with the equipment at his disposal. Even if his map is of no scientific value, it is at least of historical interest, and it was a gallant first attempt.

C22 Water on the Moon?

Look at the Moon, even with the naked eye, and you will see the broad dark patches which we always call the lunar seas (Latin, *maria*). They have been given attractive names, such as the Mare Nectaris (Sea of Nectar), the Sinus

Fracastorius, the great bay leading out of the lunar Mare Nectaris; slightly above centre. The prominent crater to the lower right is Theophilus. I took this photograph with a 15-in reflector.

Iridum (Bay of Rainbows) and the Oceanus Procellarum (Ocean of Storms). Long ago it was firmly believed that they were water filled, and that the Moon might prove to be a welcoming, life-bearing world. We know better now. The Moon has no atmosphere, and without air there can be no water. The lunar maria are bone dry, without a trace of moisture in them.

Yet have they always been so? Use a telescope, and you will soon see that they give every impression of having been fluid. Leading off the Mare Humorum, or Sea of Humours, there is a 75-mile crater, Fracastorius, whose 'seaward' wall has been so badly damaged that it can barely be traced at all, turning the crater into what is to all intents and purposes a bay. It is tempting to suggest that water may have been responsible, and this idea was current until quite recently. I well remember a meeting of the International Astronomical Union, in the mid-1960s, when Professor Harold Urey—Nobel Prize-winner, and one of the world's great lunar scientists—was trying (unsuccessfully) to convince me that the Moon had once had genuine oceans.

The rocks collected by the Apollo astronauts and the Russian sample-and-return probes put paid to this idea. The rocks contain no trace of hydrated material, and it is now generally believed that there is none at all. In this respect future lunar colonists can expect no help from the Moon. Moreover, rockets need both oxygen and hydrogen for fuel; oxygen can be taken from the rocks, but there seems to be no usable source of hydrogen.

The south polar region of the Moon. The pole is near the top of the picture, to the right; all the features are very foreshortened. Photograph by Commander H. R. Hatfield, 12-in reflector.

Or—is there? Look toward the Moon's south pole. This is one of the roughest parts of the entire surface, with large craters; some of these craters are so deep that areas of their floors are permanently in shadow, so that they never receive even a gleam of sunlight. If there is any ice on the Moon, the bottoms of these deep polar craters are the places to look for it.

Because the Moon always keeps the same face turned toward the Earth—its orbital period and its rotation period are equal, at 27.3 Earth days—the 'edges' of the disk are very foreshortened, and difficult to map without using space-craft. The polar regions have been less well charted than the rest of the lunar surface, so that what we need is a new probe which will pass directly over the poles and collect data. Several are being planned, and should be dispatched within the next few years. NASA, the American space agency, hopes to launch a lunar satellite carrying an instrument called a gamma-ray spectrometer specially for the purpose. Gamma-rays have very short wavelength; the idea is to beam these rays down on to the Moon's surface and see how they are affected, which will give a clue as to the nature of the surface material. Luckily NASA has such an instrument. It was left over from the Apollo missions, and was never sent up. (Apollo 17, the last manned lunar flight, dates back to December 1972. How long ago that seems!)

Obviously, the discovery of lunar ice would be a bonus in many ways. We are looking seriously toward a permanent base on the Moon before the year 2000, and we need all the help that Nature will give us. A lunar base can never be made totally self-supporting, but we must draw upon all the reserves we can.

Meanwhile, we can only wait. Whether or not there is any ice in those polar craters remains to be seen, but in any case the floors must be the most desolate and lonely places we can picture—bitterly cold, utterly silent, utterly lifeless, and unrelieved by any beam of sunlight for millions of years.

☾23 In the Octagon Room

A few months ago I had a telephone call from a BBC television producer who was planning a new series with the title of 'My Favourite Room'. Would I be the first subject—and if so, what room would I choose? I had no hesitation at all: I selected the Octagon Room at the Old Royal Observatory in Greenwich Park, because it was here that British astronomy really 'began' in its modern form.

The story started with the needs of the Navy. When you are far out to sea, with no land anywhere near, it is useful to know your position, and this involves making astronomical observations. Finding latitude is easy enough; all you have to do is to measure the height of the Pole Star above the horizon, and then make a slight adjustment to allow for the fact that the Pole Star is not exactly at the polar point. (Of course, southern-hemisphere observers have a harder task, because they have no suitable pole star.) Longitude is a different matter. In the 17th century the accepted method was to use the Moon as a sort of clock-hand, measuring its changing position against the starry background. This involved having a good star catalogue, and at that time even the best one available, Tycho Brahe's, was not accurate enough.

Something had to be done, and that much-maligned monarch Charles II, who was genuinely interested in science, ordered that an observatory should be built with the intention of producing a new catalogue. Typically, he paid for it by selling 'old and decayed' gunpowder to the French.

Sir Christopher Wren, who was a professional astronomer at Oxford University long before he turned to architecture, was called in to design the building. The Octagon Room is the heart of it. It really is octagonal, and

In the Octagon Room, with one of the original wooden-tube telescopes.

PROSPECTUS INTRA CAMERAM STELLATAM

In the Octagon Room.
From an old woodcut.

Old woodcut of the Royal
Greenwich Observatory,
showing the exterior of
the Octagon Room.

there are six long windows which can be opened, so that the telescopes of the period could be used from inside the room. Of course, the telescopes were very crude by our present-day standards; they had simple lenses and eyepieces and wooden tubes, but at least they had primitive micrometers, used together with the telescopes to measure small angles.

Some of the old instruments are still in the Octagon Room. For example there is a wooden telescope, eight feet long, which had to be aimed by a ladder leaned against the window; this could be moved around to any of the other windows, so that the whole of the horizon could be covered. There are some telescopes made by Dollond, who produced the first really good refractors, and also some reflectors by James Short which date back over 250 years. Other instruments are on display. What we do not have are any of the telescopes which were used by the first Astronomer Royal, John Flamsteed, who assumed office in 1675 and who did indeed produce the required star catalogue, though admittedly it took him a long time and was not published in its final form until after his death.

It is rather a sad story. Though Flamsteed was put in charge of the new observatory, there was no money to provide instruments. Flamsteed was not a rich man; he was Rector of Burstow, a little village in the country, but his salary as Astronomer Royal was a mere £90 a year, and he had to make ends meet by retaining his rectorship. Thanks to the generosity of Sir Jonas Moore, Surveyor-General of Ordnance at the Tower of London, he was able to purchase enough equipment to begin his work, and in particular to obtain an accurate clock by Thomas Tompion, but money was always a problem, and as assistant he had at first 'a silly, surly labourer' named Cuthbert, who was much more at ease in the local tavern than in the Octagon Room.

Flamsteed died in 1718, and his widow descended upon the Observatory like an east wind, removing all the instruments. Legally they were hers, and nothing could be done. The next Astronomer Royal, Edmond Halley, had to begin again.

We will never know the fate of the telescopes which Flamsteed used, but at least the Octagon Room is still there, and has been turned into a museum. It was never suitable for observing, and has not been used for anything of the kind since about 1675, but it has a marvellous atmosphere. When I go there I can just picture the old astronomers of centuries ago working away, checking on the star positions through the long windows and recording their results by means of candle-light.

 # 24 Green Stars

It is obvious that the stars are not all of the same colour. In Orion, for example, Rigel is pure white while Betelgeux is orange-red; Vega in Lyra, almost overhead from Britain during summer evenings, is steely blue; Capella in Auriga, which occupies the overhead position during winter evenings, is yellow. In the southern hemisphere we have another excellent case of contrast.

The four main stars of the Southern Cross make up a sort of kite pattern (to be candid, it is nothing like an X). Three of the members of the pattern are hot and bluish-white, but the fourth, Gamma Crucis, is orange. The difference is very marked.

Star colours are due to differences in surface temperature. Our yellow Sun has a temperature of rather below 6000 degrees Centigrade, but Betelgeux and Gamma Crucis are cooler, at between 3000 and 4000 degrees. On the other hand, Rigel measures well over 12,000 degrees, and the hottest known stars have temperatures which are much higher still.

Red stars, orange stars, white stars, yellow stars, blue stars—but what about green stars? Here we find something rather unexpected; green stars are vanishingly rare except as binary companions to brighter stars which are red.

Double stars are common in the sky, and most of them are genuine binary pairs, moving through space in company. One of these is Antares, leader of Scorpius, the Scorpion. Antares is a red supergiant, much larger and more luminous than the Sun, but considerably cooler. Its redness is striking, and indeed its name means 'the rival of Ares'—Ares being the Greek name for the war-god better known to us as Mars.

Antares has a companion of about magnitude 5½. It was probably discovered on 13 April 1819 by Burg, at Vienna, who was watching an occultation of Antares by the Moon, and saw that the reappearance was not instantaneous, so that 'perhaps Antares is a double star, and the first observed small one is so near the principal star that both, even if viewed through a good telescope, do not appear separated'. In fact the companion is not a really difficult object, and I have no problem in seeing it with my 12½-inch reflector, though it will become progressively more difficult in future years; it is a true companion, and the revolution period is 878 years, so that at the time of apparent closest

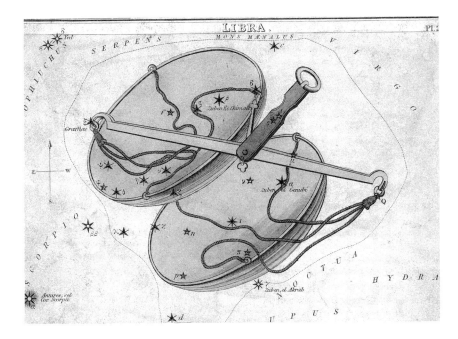

Libra, now the Balance or Scales, once the Scorpion's Claws. Beta Libræ (here 'Zuben es Chimali') is the only single star which some people say looks green.

Hardwicke Parsonage, home of the Rev. T. W. Webb. It is now a guest house.

approach (around the year 2110) it will be almost impossible to observe separately. It is often described as green, and certainly its surface is hot; the luminosity is about 50 times that of the Sun, and it is one of the very few 'normal' stars to be a definite source of detectable radio waves.

But is Antares B really green, or are we being tricked by contrast with the red primary? There are other similar cases; one is Rasalgethi or Alpha Herculis, also a red supergiant with a faint attendant which looks green. I have also heard it said that both components of the bright binary Castor, in Gemini (the Twins) have greenish casts, but to me they look pure white.

The most famous case of an allegedly green single star is that of Beta Libræ, which has the barbarous proper name of Zubenelchemale—meaning, in Arabic, 'the Northern Claw'; Libra, the Scales or Balance, was once included in Scorpius as the Scorpion's Claws. It is 120 light-years away, and has a hot surface; the luminosity is over 100 times that of the Sun. The apparent magnitude is 2.6, so that it does not shine so brightly as the Pole Star, but there have been suggestions that it has faded over the past two thousand years (personally, I am highly sceptical). It has often been described as green. A well-known American observer, W. T. Olcott, said that it 'was the only naked-eye star to be green in colour', while the Rev. T. W. Webb, author of the classic last-century book *Celestial Objects*, referred to its 'beautiful green hue'.

I have often looked at it, with the naked eye, with binoculars and with telescopes, and I have to admit that I have yet to see any greenish tinge. On the other hand I do not think that my eyes are particularly colour-sensitive, and I can only suggest that you go and look for yourself.

☾25 Doomsday, 1712

A bright comet always makes headline news, but this was also true of a comet which never appeared at all. The year was 1712; the man responsible for the whole affair was the Rev. William Whiston, a contemporary of Isaac Newton, who made a habit of predicting that the end of the world was nigh. He forecast that a brilliant comet would appear in 1712, with dire results for mankind. An account in a newspaper of the time, headed 'Ludicrous effect of the appearance of a COMET in 1712', runs as follows:

'In the year 1712, Mr. Whiston having calculated the return of a Comet, which was to make its appearance on Wednesday the 14th of Oct. at 5 in the morning, gave notice to the public accordingly, with this terrifying addition, that a total dissolution of the world by fire, was to take place on Friday morning. The reputation Mr. Whiston had long maintained in England, both as a divine and philosopher, left little doubt with the populace of the truth of his prediction. Several ludicrous events now took place. A number of persons in and about London, seized all the barges and boats they could lay their hands on in the Thames, very rationally concluding, that when the conflagration took place, there would be most

Doomsday, 1986. Shades of William Whiston! For the return of Halley's Comet, the Halley's Comet Society (which normally has no aims, objects or ambitions) organized a big charity concert at Wembley Stadium (my rôle was to provide a xylophone solo). Outside, genuine End-of-the-Worlders gathered, and were photographed. I am sorry for the poor quality of the picture. As you can imagine, it was none too easy to take!

safety on the water. A gentleman who had neglected family prayer for better than five years, informed his wife that it was his determination to resume that laudable practice the same evening; but his wife having engaged a ball at her house, persuaded her husband to put it off till they saw whether the Comet appeared or not. The South Sea Stock immediately fell 5 per cent., and the India 12; and the captain of a Dutch ship threw all his powder into the river, that the ship might not be endangered.

'The next morning, however, the Comet appeared, according to the prediction, and before noon the belief was universal, that the day of judgement was at hand. About this time 123 clergymen were ferried over to Lamberth, it was said, to petition that a short prayer might be penned and ordered, there being none in the church service on that occasion. Three maids of honor burnt their collections of novels and plays, and sent to the bookseller's to buy each of them a Bible, and Bishop Taylor's Holy Living and Dying. The run upon the bank was so prodigious that all hands were employed from morning till night in discounting notes, and handing out specie. On Thursday, considerable more than 7000 kept mistresses were legally married, in the face of several congregations. And to crown the whole farce, Sir Gilbert Heathcote, at that time head director of the Bank, issued orders to all the fire offices in London, requiring them to "keep a good look out on the Bank of England".'

It must have been fun! But there is no record that any comet appeared in 1712, so that the luckless Mr Whiston was wrong again. However, he was by no means discouraged, and timed his next Doomsday for 16 October 1736. A new comet would appear, presumably ordered specially for the occasion; the Second Coming would be timed for the previous day, and the final destruction would be in a holocaust of fire and brimstone. However, again nothing much happened. Whiston retired to Suffolk, and lived on until 1752, still confident that the end of the world would not be long postponed.

The Oldest Star Map

Star maps go back a very long way. No doubt the very earliest examples were scratched on rock walls. But it is fair to say that the first astronomers in the proper sense of the term were the Chinese and the Egyptians, who were excellent observers. For example, most of our records of ancient comets come from China, and their records are very precise. They also noted special phenomena such as eclipses, and were even able to predict them with some accuracy. The Court Astrologer was a most influential person, and his advice seems to have been carefully followed.

An exciting new discovery was made in 1990. Painted on the roof of a tomb found near the old Imperial capital of Xi'an is a star map which apparently dates back around 2100 years, making it the most ancient known. The period is that of the Han Dynasty.

The tomb was uncovered during building work at the Xi'an Jitong University. It consists of three chambers with barrel-vaulted roofs, built of mortarless bricks, and ingeniously held in position by what builders call a wedge-and-buckle design. The main chamber is gloriously decorated, with the ceiling and walls painted in pastel polychrome; the map divides the sky into twenty-eight 'lunar mansions', seven for each of the cardinal points. Each is personified by a Daoist god.

Who was responsible for it? The map matches the description given by the Court Astrologer to the Han emperor Wu Di, whose name was Sima Qian, and who lived from about 145 BC to 87 BC. He may have drawn the star map himself, and the tomb could well be his, according to the Chinese archæologists. At any rate, coins found in the tomb date from 60 BC to 87 BC, and this fits in well with Sima Qian.

We must remember, of course, that the Chinese constellations were not the same as ours. In their maps there was no Great Bear, no Orion, and of course no Southern Cross (which could not have been included in any case, because from China it never rises). In fact, Chinese star-maps are totally unlike those with which we are familiar, even though the stars themselves are the same. Constellation patterns are totally arbitrary, because the stars are at very different distances from us, and we are dealing with nothing more significant than line of sight effects.

All the same, Sima Qian's map must be fascinating. I would dearly like to see it, though I doubt if I ever will.

The observatory for the NTT (New Technology Telescope) at La Silla. I am standing in the foreground.

The Strange Nova Muscæ

Not many northern-hemisphere dwellers know much about the small constellation of Musca Australis, the Southern Fly, now known generally simply as Musca. (There was once a Northern Fly, but it has long since been dropped from our maps.) Musca Australis is in the far south, so that it never rises over Europe or the United States; it is not far from the Southern Cross, and not a great distance from the south celestial pole. There are five stars above the fourth magnitude, making up a pattern which is not hard to identify, and there are a few star-clusters within binocular range, but on the whole the constellation is not particularly distinguished. However, in 1991 a very unusual object flared up there. We call it Nova Muscæ.

Nova Muscæ was discovered in January by a Danish astronomer, Søren Brandt—not in visible light, but in X-radiation. Bodies in the sky do give off X-rays, but we cannot detect them from ground level, because they are blocked by layers in the Earth's upper atmosphere. Brandt was working with the WATCH X-ray camera carried in the Russian satellite Granat, which has been orbiting the Earth ever since December 1989. The Musca X-ray source was the second strongest in the whole of the sky, and it was unquestionably new.

The NTT (New Technology Telescope) at La Silla.

Brandt contacted astronomers at the European Southern Observatory station at La Silla, in Chile, where there is probably the most effective optical telescope yet built, the NTT or New Technology Telescope. On 14 January two astronomers there, Massimo Della Valle and Brian Jarvis, identified the X-ray source with a faint star. Over the next few nights the star brightened up somewhat, though it was still much too faint to be seen with a small telescope. It was without doubt a nova.

An ordinary nova is made up of two components, one of which is a normal star while the other is a compact object called a white dwarf, which has used up all its nuclear 'fuel' and has become amazingly small and dense. The white dwarf pulls material away from its companion, and this material collects in a disk around the dwarf; finally the situation becomes unstable, and there is a violent thermonuclear explosion, so that the star brightens up for a while before fading away again. Novæ of this kind are not uncommon, and many of them have reached naked-eye visibility, but Nova Muscæ was different. Its main energy was in X-radiation, so that it had to be exceptionally violent. Apparently the compact member of the system is not a white dwarf, but something much more exotic, made up of neutrons; it is probably no more than ten miles in diameter, but is as massive as the Sun. There is not enough spare material to create an expanding envelope, as with a normal nova, so that the outburst is caused by the heating of the disk of the neutron star itself,

The NTT or New Technology Telescope at La Silla, in Chile. Reproduced by kind permission of the European Southern Observatory.

Nova Muscæ. The left frame is a reproduction of a red-sensitive Schmidt plate (1976) and the right frame is the same field, as observed with the EMMI CCD camera on the New Technology Telescope, at La Silla, on 15 January 1991. Here the nova appears as the bright object near the centre. Now look in the left frame, and you will see the faint progenitor of the nova. Reproduced by kind permission of the ESO.

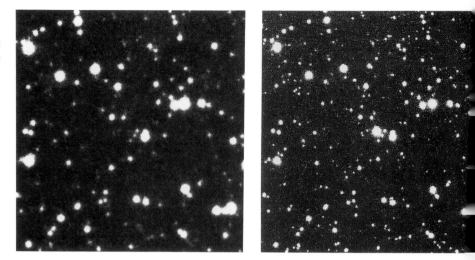

which then radiates in visible light and even more powerfully at X-ray wavelengths.

Nova Muscæ is something new in our experience, and it has caused a tremendous amount of interest, even though it has by now faded back into obscurity. It is also comforting to reflect that the research has been a joint effort by a Danish scientist working with a Danish camera on a Russian satellite, and by British and Italian astronomers based in a European observatory on Chilean soil. Nobody can now claim that science is not truly international.

☾28 Do you Want a Bet?

I am not a betting man. I am not interested in horse-racing, and I have been to only one meeting in my life. (That was in Johannesburg, more than ten years ago now; there were four races during my stay, and I backed one second and three winners, though unfortunately I put only one rand upon each.) There have, inevitably, been some astronomical wagers. It is on record that two great astronomers, Sir Arthur Eddington and Professor Findlay-Freundlich, had a dispute about a theoretical problem which was unlikely to be solved in their lifetimes; it was agreed that when the answer was found, the loser would present the winner with a new harp. Findlay-Freundlich proved to be correct, and no doubt he was duly paid!

If I did make a habit of betting, I could have had a field day in 1990 on looking at the list of bets taken by the British bookmakers, Ladbroke's. I particularly admire a Mr Simon Carreck of London, who has risked £100, at odds of 500 to 1, that the existence of UFOs—Unidentified Flying Objects—will be recognized by the British Government within the next two years. I would like to think that he might be right, mainly because I would love to meet a little green man in a flying saucer, but, rather sadly, I must regard it as unlikely. What else can we find in the Ladbroke astronomical list?

Mars is certainly the next world to be reached by man. At a recent meeting of the International Astronomical Union, a mission to the Red Planet was seriously discussed, and though it can hardly happen yet awhile it is almost on the drawing board. The general opinion was that it might take place within the next twenty years; my own view—and I know that forecasting is a dangerous business—is that 30 to 40 years is more likely, in which case I cannot hope to see it myself, though the next generation ought to do so. (Remember, the first man on the Moon—Neil Armstrong—and the first airman—Orville Wright—could have met, because their lives overlapped. I know, because I met Orville Wright and I have the honour to know Neil Armstrong.) But Ladbroke's do not believe that a Mars mission is imminent; they are offering odds of 100 to 1 against it for the coming year.

But the Moon, already reached, is a world to which we are bound to return, and here Ladbroke's are more optimistic, because they will give odds of only 50 to 1 against holidays on the lunar surface being introduced during the next decade. Bear this in mind if you have any plans for going there . . . .

All this is interesting, because Ladbroke's feel that there are events much more unlikely than the landing of an alien space-craft or a successful mission to Mars. The odds against an alien landing are a mere 100 to 1. Yet the odds against a British tennis-player winning the men's singles at Wimbledon next year are quoted at 200 to 1. I have a sad feeling that this is realistic!

I have said, correctly, that I am not inclined to take bets, but I have two wagers which are due to be paid in the foreseeable future. One of these, with my colleague John Mason, concerns the return of a comet, Swift–Tuttle, which has been seen at only one return: that of 1862. It is an important comet, because it is the parent of the main annual meteor shower, the August Perseids. The period was thought to be 120 years, so that it should have been seen again in 1982. Even allowing for an uncertainty in the period, it ought to have returned at some time during the 1980s, but it signally failed to do so. I am convinced that it has come and gone unseen; my contention is that it will not reappear during the 1990s. If it does, I lose a bottle of whisky.

My second wager, for a similar stake, is laid with an old friend of mine who ranks as one of the world's leading astrophysicists. It concerns the ideas of Dr Halton C. ('Chip') Arp, late of Mount Wilson Observatory and now working at the Max Planck Institute in Germany. Since 1963 we have known about quasars, generally believed to be the cores of very active galaxies lying at immense distances from us—many thousands of millions of light-years. Arp has reason to believe that our distance measures in the far reaches of the universe are wrong, and that the quasars are much closer and less powerful than most astronomers think. I believe he is right, and that his views will be accepted before the first day of the new century—1 January 2001 (not 1 January 2000, contrary to common belief).

We will see!

☾29 Trouble at Woomera

Many years ago, at the start of the Space Age, the main British rocket proving ground was at Woomera, in the central part of Australia. Much work was carried out, and it was confidently expected that when the space missions really began in earnest Woomera would be an important site. Unfortunately, this did not happen. The emphasis was shifted from Australia to Russia and the United States, and Woomera was left behind. Activity tailed off, and then virtually ceased. It is in a desolate part of the continent, and it was more or less left to its original owners, the Aborigines.

In 1984 the Aborigines moved back; they had been officially shifted away in the 1950s while Britain conducted nuclear tests. All seemed to be well. But then the Australian Government decided to re-open the range, not for its original purpose but to launch a series of rockets carrying experimental payloads to examine the spectrum of a spectacular object—the supernova in the Large Cloud of Magellan which blazed out unexpectedly and caused a great deal of excitement among astronomers.

As we know, a supernova is a colossal stellar explosion, ending in the destruction of a massive star. In our Galaxy we have not seen one since 1604, which was before the invention of the telescope, so that an outburst in the Large Cloud, a mere 169,000 light-years away, represented a great opportunity. The Woomera programme, which involved NASA and also West German scientists, was to lift equipment into space to monitor the fading supernova at X-ray and ultra-violet wavelengths. It meant sending rockets above the

Woomera. This picture was taken in 1956. Woomera should have become one of the great rocket grounds of the world—but, alas, it never did.

main part of the atmosphere—say 150 to 200 miles. That was when the trouble started. The Aborigines were back in the nearby Gibson Desert, and they objected strongly to the possibility of having space débris rained down on them. Their lives, they maintained, would be disrupted.

Similar problems had been encountered earlier at Kitt Peak in Arizona, where a major observatory was planned in 1950. The area comes into the reserve of the Papago Indians, and the sacred mountain Babuquivari is regarded by them as the centre of the universe, with the gods living in convenient nearby caves. After prolonged negotiations, everything was settled; the Observatory acquired the lease, and guaranteed that the sacred caves would never be disturbed. But though there are no high mountains in Australia, sacred or otherwise, the Aborigines of the Pijanjatjara Tribe were much less amenable. They did not want NASA's rockets at any price, and legally, though not in practice, they could forbid scientists access to their lands to retrieve scientific payloads and rocket boosters.

Negotiations were put in hand, but for a while neither side would compromise. There has now been a measure of agreement, and the firings went ahead, but it is understandable that the Aborigines feel much more at home with boomerangs than with supernovæ.

30 The Swedish Sun

La Palma, one of the smaller Canary Islands, has become a major astronomical centre. Here we have a whole crop of telescopes, headed by the 165-inch WHT or William Herschel Telescope, which was purely British built, and was set up in La Palma because seeing conditions there are so good. It is sited at

'The Boys': the rocks which give Los Muchachos its name.

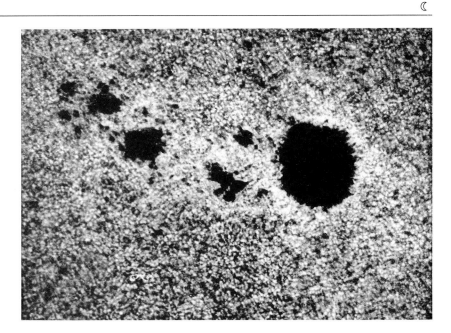

Solar granulation. This picture was taken by a telescope carried aloft in a balloon.

The Swedish solar tower at La Palma.

the top of the extinct volcano of Los Muchachos, at well over 7000 feet above sea-level, where the sky is clear and transparent for much of the time. Other telescopes include the 100-inch Isaac Newton reflector, transferred there from its old home at Herstmonceux Castle in Sussex, and the new Nordic telescope, a joint venture by Norway, Sweden, Denmark and Finland. On a visit there a little while ago I was also very interested in the Swedish solar telescope, which was actually the first large instrument to be brought into operation on Los Muchachos.

The telescope is of the 'tower' variety. That is to say, the main mirror is at the top of a tall tower; it is rotatable, catching the sunlight and sending the solar rays vertically downward, where they are reflected from a second mirror and sent into the solar laboratory at ground level. Here you can see the image of the Sun, and the various spot-groups and other features are beautifully shown. However, most of the work at the Swedish telescope is much more complicated than mere visual observation. The astronomers are concerned with all aspects of solar physics, particularly how the magnetic properties change and how the velocities of material in various regions of the Sun's atmosphere change from level to level.

Time-lapse photography is used extensively. Taking many short-exposure pictures means that you can select the best ones and use them to make up what is to all intents and purposes a movie, and speeding up by, say, five hundred times makes it possible to see the Sun 'in action', with flows of material moving in and out from the relatively dark centres of the sunspots.

Some of the most spectacular pictures are those of what we call the solar granulation. At first glance, the surface of the Sun may look bland and calm, but this is misleading; it is in a state of constant turmoil. Columns of hot gas rise up from below the main surface; each is from 450 to about 600 miles in diameter, and none lasts for more than about five minutes, but they are

The dome of the William Herschel Telescope.

ever-present. Between the granules there are darker lanes, and the time-lapse technique means that the whole panorama can be seen in motion. On the scale made possible by the use of the Swedish telescope, the granules look almost like pebbles rolling down a slope at the sea-shore.

It would be wrong to suppose that we really understand the Sun, and it may even be that we know less than we believed we did a few years ago. We are even unsure about the basic cause of the eleven-year sunspot cycle. But we are finding out more all the time, and in this field of research the Swedish telescope on Los Muchachos is playing a leading rôle.

C31 Black Hole in the Earth?

During the past few decades we have heard a great deal about black holes. Basically, a black hole is described as an area round a very dense mass of material from which absolutely nothing—not even light—can escape. It has been assumed that an old, collapsed star will produce a black hole, detectable only because of its gravitational and other effects upon objects which we can see. Of course, the star has to be very massive; our Sun, for instance, will never produce a black hole, and will end its main career as a dim white dwarf.

Black hole theories seem almost endless, and in 1990 a new one, proposed by A. P. Trofimenko of Minsk in the USSR, attracted considerable comment even though it strikes me as being frankly weird.

Trofimenko begins by suggesting that when the universe in its present form was created in a Big Bang, at least 15,000 million years ago, many tiny black

Ionian volcano: Voyager. The volcanic plume is seen rising from Io's limb. Reproduced by kind permission of NASA.

holes—less than a millimetre across—were produced. He goes on to claim that these still exist, and are to be found at the centres of all cosmical objects ranging from galaxies to minor planets. There could, he says, be some even inside the Earth. One of these mini-black holes below the terrestrial crust, a mile below ground level, it would produce a force equal to the normal strength of gravity at the Earth's surface. A black hole of this kind would warm up the region around it, causing a hot spot—and hence the localized areas of volcanic activity such as Hawaii, or White Mountain in New Zealand . . . . He even suggests that this does away with the need to assume that the Earth's core may be dense and iron-rich, and it could be of relatively low density, with a black hole in the middle.

There are sections of the Earth's crust where the density is above average, and the same is true of the Moon, where orbiting lunar satellites have detected what are called 'mascons' below some of the regular seas such as the Mare Imbrium. Again, Trofimenko assumes that these are due to sub-crustal black holes. Jupiter's moon Io is violently volcanic—same answer. And suppose a buried black hole suddenly explodes, as the great mathematician Stephen Hawking believes can happen? It would cause an outburst which would shatter material all around it—and this could account for the rings of Saturn and the other giant planets.

All of which is intriguing, but, to be honest, I think that we must dismiss it as an extreme case of making a black hole 'a remedy for everything'. I am not at all convinced that there could be a black hole lurking underneath the tennis-court in my garden. Of course, I could be wrong!

C32 X-rays from the Moon

X-ray astronomy has become an important branch of science. The Sun, predictably, is an energetic source of X-radiation, and so are many other objects in the Galaxy and beyond. But if asked to name the weakest X-ray source ever observed, even some astronomers would be perplexed. The answer is—the Moon!

In fact X-ray astronomy really began with the Moon, on 19 June 1962, with the launch of a rocket carrying spectral detectors. It was thought that some of the radiations from the Sun might hit the Moon and be re-radiated at X-ray wavelengths. That experiment failed, but during it the first true X-ray source was found; it lay in the constellation of Scorpius, and proved to be a binary star system, now known as Scorpius X-1. The Moon, as an X-ray emitter, was forgotten.

Recently it has been studied again, this time by instruments carried on a space-craft known as Rosat—more properly as the Röntgen Astronomical Satellite in honour of Wilhelm Röntgen, the German scientist who discovered X-rays as long ago as 1895. On 29 June 1990 a combined German–American team working with Rosat managed to obtain the first proper X-ray picture of the Moon.

Of course, the X-rays do not originate in the Moon itself; they come from the Sun, as was originally expected in those early experiments of almost thirty years ago. The Sun sends out a constant stream of low-energy particles, making up what we call the solar wind, and inevitably some of these particles strike the Moon. What seems to happen is that when the solar wind particles hit the lunar surface, they deposit their energy in its uppermost layer (known commonly as the regolith) and produce X-ray emission. The Moon has to all intents and purposes no magnetic field, but both it and the Earth are contained in the general interplanetary field, and some of the solar wind particles can be diverted round to the dark side of the Moon—that is to say, the area which is having its long night—and also produce X-ray emission.

It is not yet clear what we can learn about the Moon itself from these first results. There may well be a connection between the X-ray intensity and the nature of the surface in different areas, but it is too early to be at all definite. At least we do at last have a proper X-ray image of the Moon. Strange that the second most brilliant object in the sky, so far as we are concerned, should be the very weakest at these very short wavelengths!

C33 Faster than Light

The nearest star, not counting the Sun, is a dim red dwarf known as Proxima Centauri, never visible from Britain because it is too far south in the sky; it is a companion of Alpha Centauri, the brighter of the two Pointers to the

Southern Cross. Its distance from us is just over four light-years, so that we see it as it used to be four years ago. Distances of this sort seem to impose a limitation in all our attempts to send messages to beings of other Solar Systems, always assuming that they exist. It would take a radio message eleven years to reach the nearest stars which are at all like the Sun, and which might reasonably be assumed to have planetary attendants.

But suppose we could use something which travels faster than light?

Most people will say 'absurd'. According to Einstein's theory of relativity, which has been confirmed by every test so far made, nothing can outmatch the velocity of light: 186,000 miles per second. When you approach this velocity, very odd things start to happen. Your mass increases and your time-scale slows down, until at the full velocity of light your mass becomes infinite and your time stands still—which is another way of saying that it cannot be done.

Yet there could be a loophole in the form of a definite 'barrier' at the speed of light itself, leaving open the possibility of particles which always move *faster* than light but can never slow down sufficiently to reach the barrier. In 1967 a Columbia University physicist, Gerald Feinberg, even gave these particles names. He called them tachyons.

His reasoning went as follows. An ordinary particle—a bradyon—needs more and more energy as it approaches the speed of light, until it can draw upon nothing more. A tachyon needs more and more energy as it slows down, but it can never have enough energy to slow down to the speed of light. This means that tachyons (faster than light) and bradyons (slower than light) are always separated by this barrier—a kind of scientific Himalayan mountain range which cannot be crossed, but both sides of which could be inhabited.

Proof will not be easy to obtain. Way back in 1917 Richard Tolman, of California, went so far as to suggest that we might be able to use them to send messages to the past, but this seems wildly unlikely; like the White Queen in *Through the Looking-Glass*, I am prepared to believe in at least six impossible things before breakfast every morning, but even so I consider that contacting the past is one of the few things which is genuinely out of the question.

The proposal was that we might have two detectors registering the same event, one so soon after the other that we might be able to show that a faster-than-light particle had passed by. Most people share my scepticism; but if we are wrong, and tachyons do exist, it may after all be possible to send a message from the Earth to Proxima Centauri in less than four years.

Insects on the Moon

William Henry Pickering, born in Boston in 1858, was one of the best-known planetary and lunar observers of his day. He became assistant to his elder brother Edward, Director of the Harvard College Observatory, but his main interest was in the Solar System. He made careful observations of Mars, and was the first to give definite proof that the dark areas were not seas; he

*William H. Pickering.
Insects on the Moon?*

published an excellent photographic atlas of the Moon, and he discovered Phœbe, the outermost satellite of Saturn. Yet in some respects his views were rather strange, and it seems worth recalling some of them here.

For example, he believed that he had obtained observational proof that there is an appreciable atmosphere round the Moon. In particular, he studied occultations, when the Moon passes in front of a star or planet and temporarily hides it. With a stellar occultation the star snaps out abruptly, whereas it would flicker and fade briefly before vanishing if there were any lunar atmosphere (as does actually happen when Venus occults a star). A planet shows a disk, and when Pickering observed an occultation of Jupiter he recorded a dusky band crossing the planet's face. This he attributed to the effect of a thin but definite lunar atmosphere. Many other observers subsequently looked for the same sort of effect, but never found it.

Between 1919 and 1924 Pickering took his telescopes to the island of Jamaica, where the skies are very clear, and carried out a great deal of lunar and planetary work. In particular he paid particular attention to the lunar crater Eratosthenes, which lies at one end of the Apennine mountain chain; it is 38 miles in

*The lunar crater
Eratosthenes, to the
upper right, at the end of
the chain of the
Apennines. The large
crater to the bottom of
the picture is the 50-mile
Archimedes. Photograph
by Commander H. R.
Hatfield, 12-in reflector.*

diameter and 16,000 feet deep, with high, terraced walls and a prominent central mountain group. Pickering found a number of strange dark patches which showed regular variations during each lunar 'day', and although he was quite confident that vegetation tracts existed on the Moon, he suggested that the Eratosthenes patches, which moved around and did not merely spread, were better explained by swarms of insects.

This startling idea was put forward in Pickering's final paper on the subject, published in 1924. He pointed out that a lunar astronomer of a century ago would have seen similar moving patches on the plains of North America, due to herds of buffalo, and the Eratosthenes patches were about this size, though they moved more slowly—only a few feet per minute—and it was therefore reasonable to assume that the individual creatures making them up were smaller than buffalo. Although insects were considered to be the most likely answer, Pickering's paper contains the following remarkable paragraph:

> 'While this suggestion of a round of lunar life may seem a little fanciful, and the evidence on which it is founded frail, yet it is based strictly on the analogy of the migration of the fur-bearing seals of the Pribiloff Islands . . . . The distance involved is about twenty miles, and is completed in twelve days. This involves an average speed of six feet a minute, which, as we have seen, implies small animals.'

Pickering's idea was that the creatures, animal or insect, travelled regularly between their breeding grounds and the dark 'vegetation tracts' nearby. His reputation ensured that due attention was paid to his theories, but few people had much faith in them, though Pickering himself never altered his views (he died in 1938).

In fact, the idea of a very tenuous lunar atmosphere, with a ground density of about 1/10,000 that of the Earth's air at sea level, was still current well into the 1950s. It was only with the coming of the space-craft that we could be sure that the Moon is to all intents and purposes devoid of any atmosphere at all. Without water, there can be no life; Pickering's insects have, sadly, been relegated to the realm of myth. There are no caterpillars in Eratosthenes.

35 Farewell to the Castle

On Saturday, 22 April 1989, a party was held at Herstmonceux Castle in Sussex, which had for many years been the home of the Royal Greenwich Observatory. It was festive enough; the Castle was packed, and there were many people who had been associated with the Observatory for much of their careers. The Conference Room was used as a banqueting hall; in the Ballroom the band played cheerful music; in the library red and green lights flashed around the empty shelves. Yet it was not an occasion for joy. It was the last astronomical function ever to be held there: the Castle was being sold, and the Observatory re-located at Cambridge.

Old woodcut of Herstmonceux Castle— before it became the headquarters of the RGO.

The decision was made by the Science and Engineering Research Council (SERC), which controls the finances. As soon as the news was announced there was a storm of protest from astronomers all over Britain, and from abroad as well. Nobody wanted to abandon Herstmonceux, and there seemed to be no valid reason for doing so. It would not save money—a new headquarters would have to be built—and moreover the main telescopes could not economically be moved. They were not among the world's largest or best, but they were used for research, and were invaluable for the testing of new equipment. Testing of that sort could not be done at the La Palma station, simply because of the lack of available observing time.

It seemed nonsensical, but once the Civil Service has made up its mind there are obvious difficulties ahead. I became involved at the request of the Observatory; not to put too fine a point on it, I was asked to lead a campaign of resistance, because although I am a full member of the International Astronomical Union I retain my amateur status, and cannot therefore be

Herstmonceux Castle: I took this picture in 1987.

The equatorial group of telescopes at Herstmonceux, 1986. The domes still look the same, and the telescopes are still inside, but they are now mothballed.

Dome of the Isaac Newton Telescope, at Herstmonceux. The dome is now empty.

disciplined by the Civil Service or anyone else, whereas the professional astronomers were vulnerable. (Later, some disciplinary action was actually started, and was stopped only after I said bluntly that I would call a Press conference and tell everyone what was happening; for once, the Civil Service backed down.) My first act was to organize a full-scale meeting at the headquarters of the Scientific Society. It was called by me personally, under my own name, so that the SERC could not stop it, and it was attended by almost all the leading astronomers in Britain, plus a representative from the SERC. The feeling of the meeting was unanimous. Do not move: let the Observatory stay where it is. Two of the main speakers against the move to Cambridge were Professor Martin Rees and Dr David Dewhirst, of Cambridge Observatory!

At the close of the meeting I genuinely believed that we had won—but I was wrong. Despite the support from several MPs on both sides of the House, meetings in the Lords, backing from local Councils and other organizations, the SERC refused to alter its decision, and the then Minister of Science, Kenneth Baker, meekly rubber-stamped it.

During the campaign I and others circulated a memorandum. It was headed 'The Royal Greenwich Observatory should remain at its present site at Herstmonceux Castle, near Hailsham in Sussex, because:' and then followed a series of comments—too long to reproduce here in full, but a few of which may be quoted.

(1) None of the investigations carried out by various committees over the past five years have been able to demonstrate any financial advantages in moving from Herstmonceux.

(2) Moving the RGO would drastically impair its close contact and proximity to the majority of UK astronomical groups.

(3) Moving the RGO would sever the close link between it and the important Sussex University Astronomy Centre.

(4) The long-standing research team at the RGO would be broken up.

(5) There would be severe disruption of activities in La Palma, partly because of the loss of morale among RGO staff members.

(6) The move would result in the permanent loss of many key personnel in high technology areas.

(7) Any move would involve the loss of the telescopes set up at the Equatorial Group in Herstmonceux.

(8) Merging with a university would mean that within a few years the RGO would lose its separate identity.

(9) The SERC had listed no shortcomings in Herstmonceux as a base for the RGO, and had identified nothing that could be done better at a different site.

(10) The Library, one of the world's best, would be broken up.

(11) The Exhibition would be closed; there would be no room for it at Cambridge.

(12) The educational facilities, unique to Herstmonceux, would also be lost.

The memorandum ended by pointing out that Herstmonceux was by far the best site in purely physical terms; there was plenty of room for expansion, even if it were wished to hold more functions there—even to increasing the use of the Castle as a major conference centre.

None of these arguments impressed the SERC. The Castle was abandoned, and sold at a surprisingly low price to a property speculator, who has since

Inauguration of the 'new' Royal Greenwich Observatory at Cambridge, by Robert Jackson, MP, after the move from Herstmonceux. Left to right: Professor Alec Boksenberg, Director of the RGO; myself, and Mr Jackson. To say that I felt as cheerful as I looked would force me to be economical with the truth. Photo: courtesy of the Rutherford–Appleton Laboratory.

done nothing with it—it has been several times offered for re-sale. The telescopes remain on the site, mothballed and falling into disrepair. The Observatory is now located in an office block in Cambridge, minus its library, its archives, its educational facilities, its telescopes, and most of its original staff.

Yes; that farewell party in April 1989 was a sad occasion indeed.

☾36 Black Holes and Blue Stragglers

Star clusters are of two definite types. Loose or open clusters, such as the Pleiades, contain from a few dozen to a few hundred stars, and have no particular shape. Globular clusters are quite different; they are huge, spherical systems, lying near the edge of the main Galaxy, and may include more than a million stars each. Most of them are thousands of light-years away, so that they appear faint. Only three are clearly visible with the naked eye: the Hercules cluster (Messier 13) in the northern hemisphere, Omega Centauri and 47 Tucanæ in the southern.

Of these Omega Centauri is much the brightest, but for sheer beauty I always think that the palm must be given to 47 Tucanæ, which lies very near the Small Cloud of Magellan and looks almost as though it were part of it—though in fact the distance of 47 Tucanæ is only 15,000 light-years, while the Small Cloud is an external galaxy at more than 150,000 light-years. Near the core of a globular cluster the stars appear to be very crowded together, and it is not easy to distinguish them separately. The best views so far obtained of the centre of 47 Tucanæ have been taken by the Hubble Space Telescope, and have led to some very independent conclusions.

Globular clusters are very old. They date back to the time when the universe, in its present form, was young—perhaps 14 to 15 thousand million years ago—and this means that their leading stars ought to be red, because a red star has used up most of its nuclear 'fuel' and is senile. But as long ago as 1953 the American astronomer Allan Sandage found that the core of 47 Tucanæ also contained some blue stars. They should not have been there, so they were nicknamed Blue Stragglers. In 1991 the Hubble Telescope, with its great resolving power, identified twenty-one of these stragglers in the heart of the globular cluster. What makes them blue?

As we know, a star is created out of the dust and gas in a nebula, heats up, shines for a long time as a hot white or blue star, swells out to become a red giant, and then dies, either by exploding as a supernova or (if less massive) collapsing into the white dwarf state. The stars in 47 Tucanæ should have had ample time to do this; yet the blue stragglers are there, and we have to explain them somehow.

It is not likely that they are young, because globular clusters used up all their star-forming material long ago. Neither is it likely that they have mixed

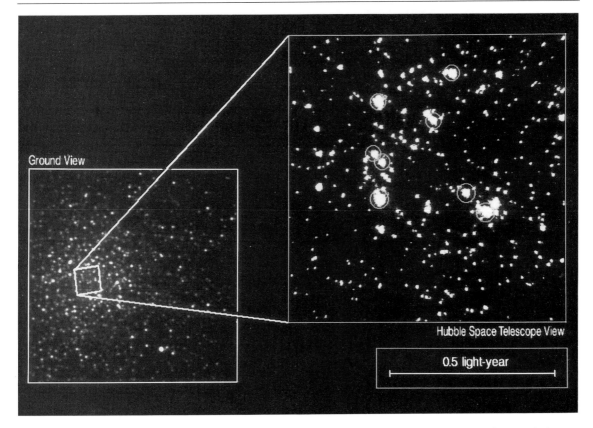

Blue Stragglers. These are two pictures of the globular cluster 47 Tucanæ. On the left, a ground-based image, taken with a CCD on the ESO 2.2-metre telescope at La Silla in Chile; the photograph is in blue light, with a resolution of 0".8 and a limiting magnitude of 16. On the right, the boxed region as photographed with the ESA's Faint Object Camera on the Hubble Space Telescope; several hundreds of stars are seen. The Blue Stragglers are encircled. Resolution, 0".1. Reproduced by kind permission of ESA and NASA.

their hydrogen 'fuel' so thoroughly that they could shine for much longer than is usual; they would need to improve their efficiency by over one hundred per cent. So the best answer may be that proposed in 1964 by W. H. McCrea and Fred Hoyle. They pointed out that in the core of a globular cluster the stars are packed relatively closely (if you lived on a planet there, you would see more than a million naked-eye stars on a clear night, as against our modest 3,500), and the nearest star would be no more than twenty times the distance of Neptune from our Sun. Moreover, the stars would pass by each other at a relative speed of no more than about 10,000 miles per hour, which is not very much bearing in mind that our Sun moves round the centre of the Galaxy at the rate of about 40,000 m.p.h. Therefore, stars might 'capture' each other to form binary systems. The less massive star would siphon fresh hydrogen from its quicker-evolving companion, and would heat up, turning blue again. With a direct collision, the two stars might actually merge.

This seems feasible. The merged stars would be of more than average mass, and would be expected to collect near the core of the cluster, which is where we find the blue stragglers of 47 Tucanæ. It all fits.

On the other hand, the Hubble Telescope has found no sign of the presence of a black hole in any globular cluster; previously it had been thought that each globular might have a black hole in its heart. So far as we can judge at the moment, the message from Hubble is: Black holes 'out', blue stragglers 'in'!

Globular cluster. Near the core of the cluster, the stars are relatively densely packed together.

The Great Wall— From the Moon

Quite recently I was talking to Commander Eugene Cernan, commander of Apollo 17, who has the distinction of being (so far) the last man on the Moon. When he followed Harrison Schmitt into the lunar module, in December 1972, and blasted back into orbit, the first phase of Man's exploration of the lunar world was over.

I remember asking him to tell me what was his most vivid impression. He replied that it was the view of the Earth, seen from a distance of a quarter of a million miles. I was not entirely surprised, because other Apollo astronauts had said much the same, but it started a chain of thought in my mind. Over that range, how much detail on Earth can you see with the naked eye?

There has been an oft-repeated claim that the one man-made object visible is the Great Wall of China. Of course, the outlines of the seas and continents are clear enough when they are not hidden by cloud, but artifacts come into a different category. Certainly the Great Wall could be a candidate, because

The Earth from space: four pictures taken by the Galileo space-craft on 11 December 1990, at ranges from 1.2 to 1.7 million miles. Antarctica is shown at the bottom. Reproduced by kind permission of NASA.

of its enormous length, and it is well known that the human eye can pick out 'long' objects much more easily than short ones.

However, the claim has now been investigated by a well-known photographic expert, H. J. P. Arnold, who has made a special study of this sort of problem, and is also a skilful astronomer. He has come to the conclusion that the claim is completely out of court. I quote:

'The reality is that apart from a small renovated section near Pekin, the Wall is a frequently broken-down edifice. Far from being seen from space, it can often scarcely be seen on the ground. And though the human eye is extremely good at distinguishing lines from a great distance, seeing the Wall from space is a physical impossibility.'

To confirm this, we have the testimony of Neil Armstrong, who has said definitely that from the Moon the Wall is not visible. And Jim Lovell, who flew in Apollos 8 and 13 and made very careful observations, says that the claim is absurd. On the last occasion when I met Jim Irwin, who went to the Moon

The Great Wall of China, not visible from the Moon! This picture was taken from terra firma in 1991, by Colin Ronan.

in Apollo 15, he also said that seeing the Wall was out of the question. It is difficult to see from space at any time, and photographs from unmanned probes show that though its route is shown by sand blown on its windward side, the Wall itself is not on view. End of yet another legend!

On the other hand, you can see stars in the daytime if you are on the Moon, provided that you shield your eyes from the glare of the lunar rocks. That was something else that I learned from the astronauts, and it must be a fascinating sight, but the Great Wall of China—no.

38 The Doomed Satellite

The red planet Mars has always been regarded as the most earthlike of our neighbours. True, it is much smaller and colder than the Earth, and its thin atmosphere is quite useless so far as we are concerned, but at least it is much less unfriendly than Venus or Mercury, the other members of the inner group of planets. Plans for sending expeditions there are already being made.

There are two satellites, Phobos and Deimos, both discovered by Asaph Hall as long ago as 1877. Both are small, so that as seen from Earth they are faint; I can just about see them with my 15-inch reflector when Mars is well placed, but not easily. They are irregular in shape; Phobos has a mean diameter of 14 miles, Deimos only 8. Also, they are close-in, and Phobos moves round the planet at only about the same distance as London is from the Gulf. Another peculiarity is that Phobos completes one orbit in 7½ hours, while Mars takes 24½ hours to make one full turn; in fact, so far as Phobos is concerned the 'month' is shorter than the 'day'. To a Martian observer, then, Phobos would rise in the west and set in the east, crossing the sky in only 4½ hours, while the interval between successive risings would be little more than 11 hours. Neither satellite would be of much help in lighting up the landscape during night-time. Both move more or less in the plane of the planet's equator, so that from the Martian polar regions they would not be visible at all.

In 1945 B. P. Sharpless, in America, made a careful study of the movements of Phobos, and found that it was gradually speeding-up in its path, so that it would eventually spiral downward and crash into Mars. This led the eminent Russian astronomer Iosif Shklovskii to make the extraordinary suggestion that Phobos was being 'braked' by the thin Martian air at that altitude, so that it had to be hollow, and was a space-station launched by the local inhabitants . . . The Soviet Academy of Sciences was not impressed, and years later, when I asked Shklovskii about it, he told me that his suggestion had been nothing more than a practical joke. By then, however, the American space-craft had taken close-range pictures of it!

New studies by Andrew Sinclair of the Royal Greenwich Observatory have confirmed that Phobos really is spiralling downward at the rate of about 60 feet per century. This means that it will crash on to Mars in 40,000,000 years from now, while the orbital period will decrease from its present 7½ hours to only 1 hour 40 minutes just before it lands. The other satellite, Deimos, is further out, and seems to be safe.

There is an important conclusion to be drawn from this. Forty million years is not a long time by astronomical standards, and it seems likely that Phobos and Deimos are not true satellites of Mars. Further from the Sun we come to the belt of asteroids, small worlds which may be regarded as planetary débris. Probably, therefore, Phobos is an ex-asteroid which was captured by Mars in the remote past. Otherwise it could not have lasted until now, bearing in mind that the planets are well over 4,000 million years old. It is in no danger of imminent destruction, but we know that eventually it is doomed.

39 The Hunt for Phobos 1

It is a curious fact that although Russian space scientists have had great success with their missions to Venus, they have had no luck at all with Mars, which should be a much easier target. The latest failure came with two probes which held out hopes of being exceptionally interesting. They were called Phobos 1 and Phobos 2.

They were launched in July 1988, not mainly to study Mars itself, but to concentrate upon Phobos, the larger of the two tiny satellites—an irregularly shaped, dark object with a longest diameter of 17 miles. The two Soviet space-craft were similar, though the second one carried equipment which was intended to 'hook on' to the satellite. There was even a landing section which would, it was hoped, hop around the surface of Phobos like a frog.

The launches themselves went well: Phobos 1 started its journey on 7 July, and Phobos 2 five days later. They should have reached Mars in January

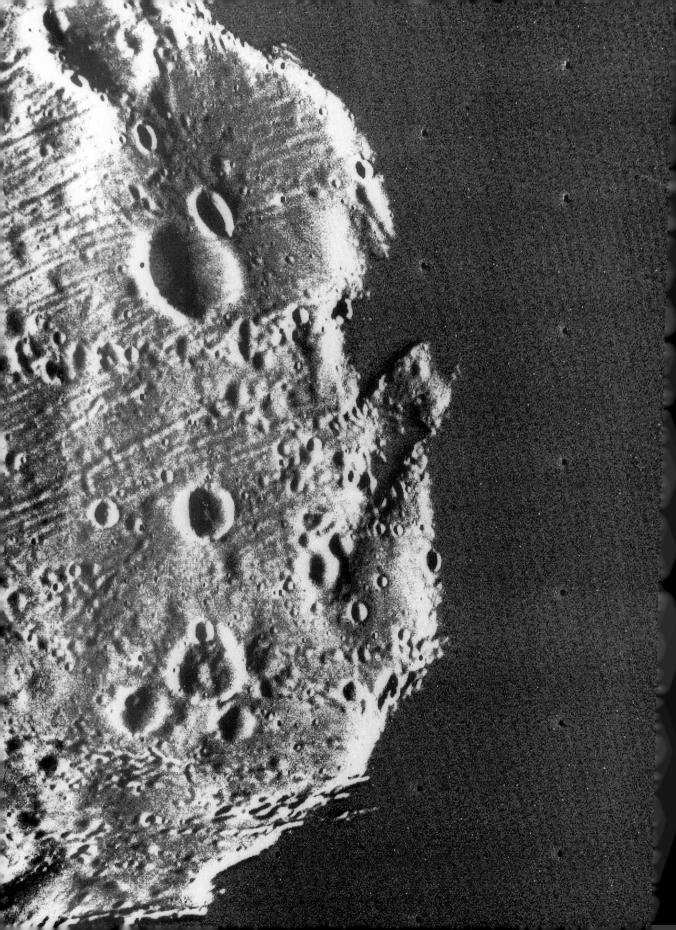

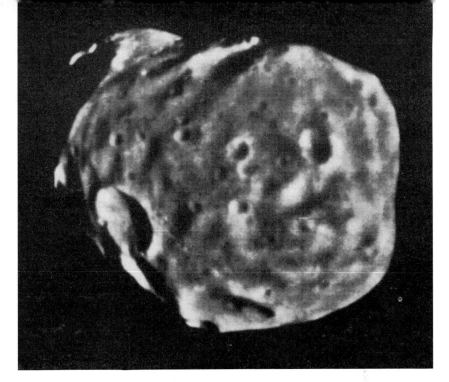

Phobos, from the Phobos 2 space-craft on 21 February 1989. Range, 273 miles.

Phobos, imaged on 18 September 1976 by Viking 2 from a range of 545 miles. Features down to 130 feet resolution are shown, Phobos' north pole is near the upper left. (Facing page.)

1989, making their Phobos encounters in the following April. But then, on 28–9 August 1988, came a disastrous error by one of the Soviet ground controllers. A faulty command was sent to Phobos 1, and the computer did not override it. The space-craft automatically shut down its rocket thrusters and veered over, so that its solar panels swung out of alignment with the Sun, and the electrical and electronic systems lost all power. Phobos 1 went out of contact.

Roald Sagdeev, who was for many years in charge of the Russian space programme, was scathing. So far as Phobos 1 was concerned, he said, 'its on-board computer chips are probably damaged beyond repair because of the loss of power. I see virtually no hope of getting the space-craft working again.' Neither was he happy about Phobos 2, because the same ground control team was dealing with it.

Running repairs in space are not exactly easy if they have to be carried out over a distance of millions of miles, but it was thought that there might be a chance of restoring the situation if only Phobos 1 could be located. So on 21 September 1988 the Russians sent a request to the La Silla Observatory in Chile. This is the main station of the European Southern Observatory, and is multi-national, though, regretfully, Britain is not directly involved. Conditions there are as good as they are anywhere in the world, and the Soviet request was straightforward: Would La Silla try to photograph Phobos 1? If it were possible to obtain a sequence of images, of course only as points of light, a knowledge of its exact position and even rotational status might give the Russians a chance of re-activating it.

La Silla promptly agreed. (Would that the collaboration so evident in astronomy extended to all other sciences!) The first night, 22 September,

The Danish 60-inch telescope at La Silla, used in the hunt for Phobos 1.

was cloudy, but on the nights of 2, 3 and 4 October a careful search was made, using the latest electronic equipment on the 60-inch Danish telescope. Fifteen ten-minute exposures were made. On 3 and 4 October the largest telescope then operating at La Silla, the 158-inch reflector, was also used for the hunt by two visiting astronomers, D. Hatzidimitriou and C. A. Collins. On one exposure, the 158-inch telescope was set to make a direct exposure of the field in which Phobos 1 was expected to be. On another, the telescope was set to 'follow the space-craft' in its predicted motion, so that the stars would be drawn out into trails. In each case the limiting magnitude was below 25.

Alas! there was no sign of Phobos 1. The plates were intensively studied at La Silla, by H. U. Nørgaard-Nielsen and H. Pederson, and if there had been a trace of the elusive space-craft it could not have been overlooked; but it simply was not there.

La Silla did not give up, and on 9–10 October two more visiting astronomers, G. Suchail and Y. Mellier, made a final effort. They took four plates, extending over a period of 36 minutes, but again Phobos 1 declined to appear. That, sadly, was as much as could be done. Either Phobos 1 was outside the expected field of view (did one of its on-board rockets fire after contact was lost?) or else it was positioned so unluckily that the sunlight reflected off its surface in our direction was too faint to be recorded.

Phobos 1 was never found, but the whole episode underlined, yet again, the way in which astronomers are ready to help each other. If anyone could have located the missing probe, it would have been the observers at La Silla.

Incidentally, Phobos 2 did remain in contact until it reached Mars, but 'went silent' before the Phobos experiments could begin, doubtless because of an on-board fault. So the Russians have yet to achieve real success with the Red Planet.

40 Comets and Gamma-rays

In 1991 a new artificial satellite, GRO—the Gamma-Ray or 'Compton' Observatory—was launched from the Space Shuttle. It was put into a path which carried it round the Earth at 280 miles above ground level, and was designed to operate for at least two years. It was the heaviest NASA load ever deployed by the Shuttle; its weight was 35,000 pounds.

Gamma-rays are ultra-short, and have to be studied by space research methods simply because they cannot penetrate to the Earth's surface. Various discrete sources have been found. Pulsars (neutron stars) send out gamma radiation; so do quasars and some active galaxies, while the Milky Way shows a general gamma-ray glow which is not understood. There are also 'bursters', which are particularly hard to study because they cannot be predicted and do not last for long. For example, one burster recorded in 1979 lasted for only a tenth of a second, but during that time it sent out more gamma-ray energy than our Sun will do over the next thousand years.

All sorts of theories have been proposed to explain the gamma-ray bursters. The latest comes from a leading Russian astronomer, Roald Sagdeev, who was in charge of the Soviet missions to Halley's Comet in 1986. His solution is unexpected, as it involves not only neutron stars but also comets.

Neutron stars, as we know, are the remnants of supernovæ; take a teaspoon of neutron star material, and it will weigh at least a thousand million tons. According to Sagdeev, gamma-ray bursts are caused when a comet passes

Gamma-ray telescope. This strange instrument, at Mount Hopkins in Arizona, is designed to detect some gamma-rays from space—even though most of these ultra-short radiations are blocked by the Earth's atmosphere.

through the intense magnetic field of a neutron star. Strong electric currents are set up in the comet, destroying it; the matter from the dead comet then short-circuits the magnetic field, and a burst of gamma-rays is the result.

Comets have always been regarded as very flimsy things, of very small mass, so could they possibly have this kind of effect? Sagdeev thinks so. When the various space-craft went closer to Halley's Comet they found that the nucleus of the comet was dark, not bright as had been expected. Up to that time, astronomers had estimated the masses of comets from their apparent brightness, on the assumption that they were good reflectors of sunlight, but we now know that this is not so. It follows that comets may be up to sixty times more massive than had been believed.

It is also thought that comets in our Solar System come from a whole cloud of icy objects at a distance of around one light-year from the Sun. (One light-year is equal to rather less than six million million miles.) In the sky we know of gamma-ray sources which have burst out more than once, and Sagdeev suggests that this is because the neutron star has a similar comet cloud around

Gamma-ray equipment being built at the University of Durham. This will be set up at Narrabri in Australia.

it; we pick up the gamma radiation when a new comet plunges in from the cloud and is destroyed.

We cannot pretend to be sure, but at least this is a possibility. It could well be that a burst of gamma-rays marks the death of a luckless comet as it is destroyed by the magnetic field of a neutron star.

41 Astronomical Tables— £104,000!

Would you spend £104,000 on a set of astronomical tables? I admit that I would not, even if I were a multi-millionaire; but it happened in April 1988 at a sale at Christie's, the London auctioneers. The tables were compiled in the 13th century, and are known as the Ilkhanic Tables. They were master-minded by a Persian astronomer, Nasir-al-Din Tusi, better remembered by us as Nasireddin.

Ptolemy, the last great astronomer of classical times, died about the year 180. Subsequently there was a long period of stagnation, but then came the Arabs, who were essentially astrologers and who needed good star catalogues and tables of planetary movements to enable them to draw up their horoscopes. Next there were the Persians. Nasireddin was born in 1201, and his genius was recognized by the Persian rulers. Hülugü Khan, grandson of Genghis Khan, made him Vizier, and required him to build a major observatory at Marahgah. By 1259 it was complete. Of course it had no telescopes (telescopes did not come on the scene for another 350 years or so), but it had excellent measuring instruments, and Nasireddin assembled an impressive collection of scholars, including Omar Khayyám and a Chinese astronomer named Fao Mur-Ji. Theoretical and observational work was carried out, and the Ilkhanic Tables were so good that they were used as a basis for several centuries afterwards.

At Marahgah there was a library of over 400,000 books, and at that time Persia was the leading astronomical centre of the world. Later, in Samarkand, a great observatory was set up by Ulugh Beigh, and this really marked the climax of astronomy in the area. Unfortunately for himself, Ulugh Beigh was a firm believer in astrology, and banished his eldest son, Abdallatif, upon astrological advice; it was claimed that the boy was destined to kill him. Abdallatif returned, raised an army, invaded his father's realm and had Ulugh Beigh murdered. That was one prediction which came true, and it also marked the end of a great school of astronomy. Henceforth, the scene shifted to Europe.

But let us not forget Nasireddin, who was regarded as one of the 'Great Wisdoms' of his time. He died in 1274, more than seven centuries ago now, but in his day he was pre-eminent, and he will not be forgotten even by those who would think twice about paying over a hundred thousand pounds for a set of his astronomical tables.

Martian Mud

My memory goes back to the time when Mars was regarded as a life-bearing world. The 'canals' drawn by astronomers such as Percival Lowell had been discounted, but it was thought that there must be plenty of vegetation, albeit of lowly type. The space-craft results have given us a different picture, but we cannot yet be absolutely sure that Mars has always been sterile. We have photographs of features which are almost certainly old riverbeds, so that there must once have been running water on the surface even though the atmospheric pressure is too low for water to exist there now.

There seems little doubt that Mars and the Earth are of about the same age—around four and a half thousand million years—and that both were formed from a cloud of material associated with the youthful Sun. According to a recent theory by Anastasia Kanavarioti and Rocco Mancinelli, working in America, Mars and the Earth started to evolve along the same lines. They produced similar atmospheres; they had seas and rivers; and as we now know that life on Earth began at a much earlier stage than used to be thought, there is no reason to doubt that the same applied to Mars. Around three and a half thousand million years ago, then, both planets were life-bearing. But then the situation changed, solely because Mars is much less massive than

Mars, photographed with the wide field and planetary camera of the Hubble Space Telescope. The dark triangular marking is known as the Syrtis Major. Reproduced by kind permission of NASA.

the Earth. Its weak gravitational pull was unable to hold down a substantial atmosphere, so that the Martian air leaked away into space, and life died out. Volcanic activity ceased, and Mars became the dry, geologically quiet world of today.

However, we might still find traces of life. Kanavarioti and Mancinelli maintain that these traces may lie beneath the surface as broken-down products of amino acids, which are essential components of proteins and are characteristic molecules of all living organisms. Amino acids can remain intact for very long periods, and if they ever existed on Mars we could still find traces of them.

Another new theory, due this time to Eric Christiansen of the Brigham Young University in America, suggests that we might find signs of life in mud below the planet's surface. The Viking pictures, taken from close range, indicate the presence of channelled flows of volcanic débris known as lahars, not unlike those associated with Mount St Helens in Washington State. Lahars are wet mixtures of water and solids; those from Mount St Helens were from 22 to 36 per cent water. Christiansen believes that in that part of Mars known as Elysium—a volcanic region—there are channelled deposits, now dry, which are of the same type. They are branched, and this implies that seepage valleys were cut as water was expelled to the surface.

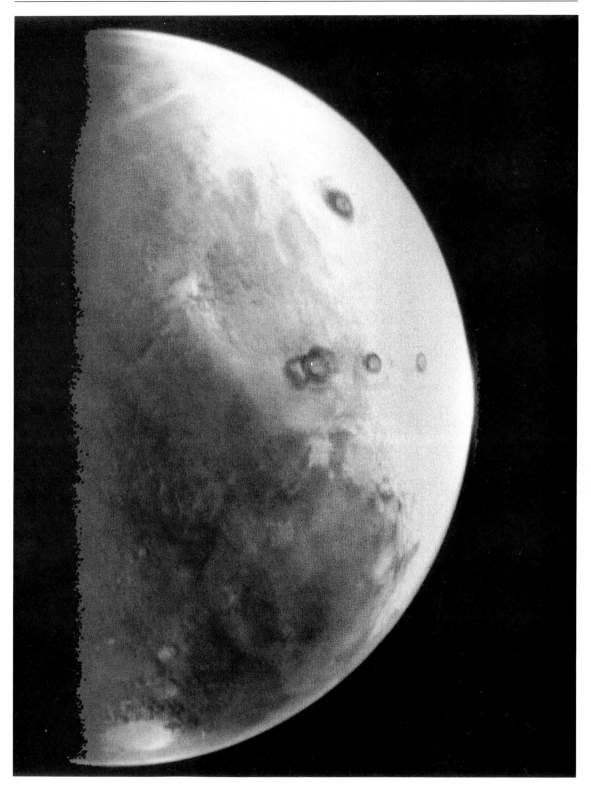

Mars: taken by Viking 1 before arrival. The huge volcano above centre, with the white cloud coming from it, is Olympus Mons. Polar deposit is seen at the bottom of the picture. Mars today is a lifeless world—so we believe!—but was it always so? Reproduced by kind permission of NASA. (Facing page.)

Martian islands. This Viking 1 picture (23 June 1976) shows 'islands' in the channel of the Ares Valley on Mars, not far from Chryse, where the Viking 1 lander subsequently touched down. Markings on the channel floor show the direction of the water-flow, and water and wind have etched out the layers of rock which form the islands. The range from which the picture was taken was 1070 miles. Reproduced by kind permission of NASA.

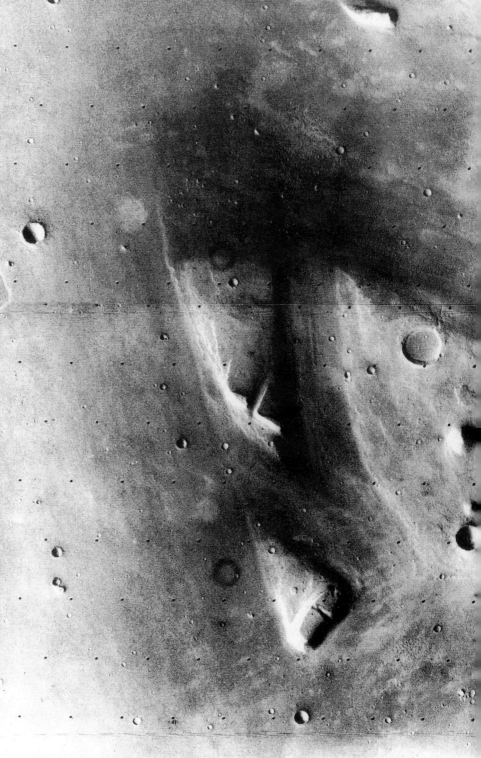

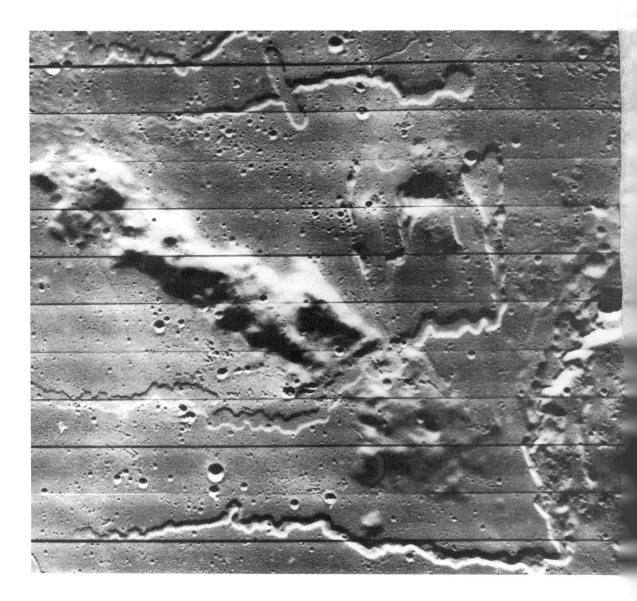

The riverbeds of Mars. This Viking space-craft picture shows what certainly seem to be dried-up riverbeds on Mars. Reproduced by kind permission of NASA.

If this is so, then long ago, when the Martian volcanoes were active, mud flows were produced, now showing up as the channels which we see. As we have known for some time, there can be no liquid water on the surface today, but suppose that deep down, out of our sight, there is mud?

We must be very careful not to jump to conclusions, but there seems no doubt at all that liquid water once existed, and it follows that at that period Mars was much more fertile than it is now. If so, then life could have developed. It is not likely that there were higher forms—the 'Martians' belong to science fiction—but primitive creatures are more possible. We will find out only when we can bring back samples from Mars, both from the surface and from below, to analyse in our laboratories.

I am still rather sceptical about the chances of finding Martian fossils, but I would not rule it out completely. Certainly the discovery that life once existed on the Red Planet would be dramatic indeed.

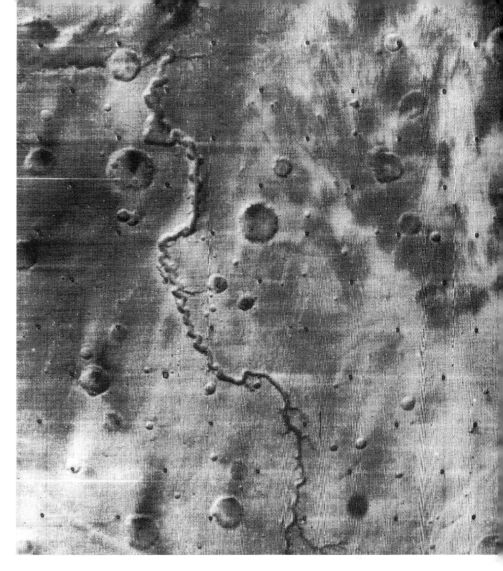

Martian arroyo. This is apparently similar to the watercut gullies such as those in the SE United States; it is 250 miles long and 3 to 3½ miles wide. The centre of the image is at lat. 29° S, long. 40° W. The picture was taken from Mariner 9 from a range of 1033 miles, in February 1976. Reproduced by kind permission of NASA.

43 Protesters and Plutonium

On 18 October 1989 I went to Cape Canaveral, in Florida, to watch the launch of a space-craft aimed at Jupiter. It was named in honour of the first telescopic observer, Galileo, who had turned his tiny 'optick tube' toward the Giant Planet in January 1610, and had made the first systematic studies of the four large satellites which have proved to be such fascinating bodies. When the probe is nearing Jupiter, five times as far away from the Sun as we are, it cannot use solar power for its instruments. Instead, it carries a small radioactive generator powered by plutonium—and this was the cause of the trouble.

Plutonium is highly dangerous, because even a tiny amount breathed into the lungs will cause cancer. There was a suggestion that if the launcher—the

Model of the Galileo space-craft, at the Jet Propulsion Laboratory (1991). Whether or not Galileo will succeed in obtaining results from Jupiter remains to be seen, but at least it has sent back some impressive pictures of the Earth and Moon.

Shuttle *Atlantis*—exploded in the way that *Challenger* did, plutonium might be scattered all over Florida, and cause numerous deaths. An organization calling itself the Florida Coalition for Peace and Justice, spearheaded by a Mr Bruce Gagnon, took the lead, and enlisted students from the Rollins University, who vowed to sit on the site and prevent the launch. They took their case to the Federal Court, and the decision was left to the local judge, Oliver Gaskin.

The arguments were forcefully presented. Initially, the protesters claimed that NASA had no experience of this sort of launch (though, of course, they had). Judge Gaskin replied that Columbus didn't have much experience either. Mr Rogers, for the protesters, countered this with 'Columbus wasn't going to contaminate the world'. In the end, Judge Gaskin ruled that the risks were so slight that there was no reason to interfere; Galileo took off safely, and the Rollins students did not sit on the site. In fact we saw absolutely no trace of them, though security men were very much on the alert.

But is there any real reason for alarm?

First, the chances of an explosion are slight. Even if *Atlantis* had blown up, the plutonium would have fallen to the ground in solid lumps contained within protective shields, and it is no danger unless breathed directly into the lungs. Therefore, the only genuine risk would be if Galileo re-entered the atmosphere and burned up. Certainly its orbit is complicated; it cannot

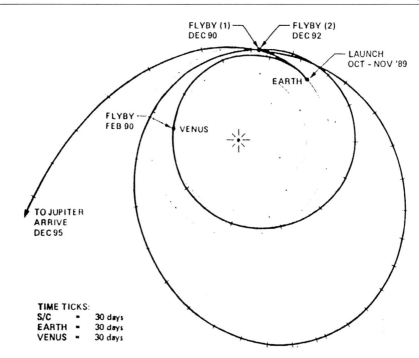

Orbit of Galileo.

go directly to Jupiter, as the Pioneers and Voyagers did, because its launcher was not sufficiently powerful. It makes one fly-by of Venus (February 1990) and two of the Earth (December 1990 and December 1992) before starting on the final leg of its journey, taking it to a rendezvous with Jupiter in late 1995. During the second and closer Earth pass it will skim by us at only 200 miles, and there is, I suppose, just a chance that it could go off-course and plunge back into the atmosphere.

Careful studies have been made, and the odds against any accidental re-entry have been given as about one in two million, which must be regarded as acceptable. Moreover, Galileo should remain under control, so that any deviation from the planned path can be remedied well ahead of time. Therefore, it seems that we have nothing to fear from Galileo, though it is true that we need to watch the space environment in general, and military satellites carrying radioactive material are emphatically not to be encouraged.

44 The Story of Stellaland

Have you ever heard of the Republic of Stellaland? I admit that I had not, until I went recently to stay with my friend Dr Hamish Campbell, at Durban in South Africa. Dr Campbell is one of Durban's leading medical doctors. He is also a well-known amateur astronomer, and he is an expert stamp collector too. It was when he showed me a Stellaland stamp that I first encountered this strange, short-lived republic.

The story began in 1882, when there was political unrest in the Transvaal. This was nothing new, of course, but it was accompanied by an economic slump, and many of the inhabitants became so restive that they decided to move elsewhere. There was a Boer trek northward, into the area now occupied by Botswana, and two small republics appeared, declaring their independence. One, known as Goschen, had its capital at Mafeking. The other, Stellaland, was centred round Vryburg.

The name was chosen simply because at the time a particularly brilliant comet was in view, and it seemed as though it might be a favourable omen. The comet (one of two brilliant visitors seen in that year) had been discovered in September, and by the middle of the month was easily visible in broad daylight. It was close to the Sun on the 17th, and Dr Elkin, from the Cape of Good Hope, described the nucleus as 'scarcely fainter than the Sun's limb'. During the night it cast shadows whenever it was above the horizon, and it had a dense, white, sharply-bordered tail. It was without doubt one of the most spectacular comets ever seen, and the leaders of the new republic were suitably impressed.

Nobody seems to have much idea of who really organized and controlled Stellaland, but it is clear that the infant republic did no harm to anyone. It issued stamps which were countersigned by one J. P. Minaar, who was apparently Stellaland's Treasurer-General and is the only official whose name I have been able to identify.

Neither Stellaland nor Goschen was recognized by any important nation, but normally this would not have mattered, and the two would have been allowed to continue peacefully. The problem lay in their position. Cecil Rhodes was organizing the Cape to Cairo Railway, and both Goschen and Stellaland were in the way.

Negotiations seem to have been tried. The Imperial Deputy Commissioner, the Rev. John Mackenzie, offended the Stellalanders by ignoring their rights; Rhodes tried to placate them by guaranteeing their land titles, but the writing was on the wall. The London Convention decreed that the whole area should be annexed, and the Cape authorities dispatched a force under Sir Charles Warren to cope with the situation. When Warren's men arrived, Stellaland ceased to exist. Not a shot was fired, and the little republic was officially ended on 7 February 1885.

All in all, this seems rather a pity. Certainly Stellaland had a romantic name, and I suppose there is a chance that it could have survived, at least for a while, if it had been in a different place. It deserves to be remembered—if only because despite its brief life and its lack of importance, it will presumably remain the only independent republic to be named after a comet.

The Future Moon

Quite recently I found an old periodical—Newnes' *Practical Mechanics*—dated September 1936. In it was an article by Sir James Jeans, one of this century's greatest astronomers. It was entitled 'How the Moon will Disintegrate', and it seems to be worth quoting—bearing in mind that our views have changed out of all recognition since 1936!

First, Jeans describes the way in which the Moon causes the ocean tides. Tidal undulation acts as a brake on the Earth's rotation, and the days are getting longer (which is quite true, though by only a tiny fraction of a second per century). Now for Jeans' words:

> 'The tides will continue to act in this manner until the Earth and Moon are rotating and revolving in complete unison. . . . The Earth will continually turn the same face to the Moon, so that the inhabitants of one of the hemispheres of the Earth will never see the Moon at all, whilst the other side will be lighted by it every night. By this time the length of the day and month will be identical, each being equal, to about 47 of our present days. After this, tidal friction will no longer operate in the sense of driving the Moon further away from the Earth. The joint effect of solar and lunar tides will be to slow down the Earth's rotation still further, the Moon at the same time lessening its distance from the Earth.

> 'The Moon will gradually come nearer to the Earth until its distance is but little more than 12,000 miles instead of an average of 240,000 miles as at present. . . . As observed from the Earth, the Moon would always appear in the same place in the sky, immense and menacing. . . . A catastrophe would now be pending. When the Moon has been dragged down to within about 12,000 miles of the Earth, the tides raised by the Earth in the solid body of the Moon will shatter the latter into fragments which will form tiny satellites revolving round the Earth. Numerous fragments will fall as giant meteorites, raised to incandescence in their terrific speed through the Earth's atmosphere. . . . A large portion of the Moon would ultimately be resolved into dust, which would be spread out as a vast ring encircling the Earth in the same way as the particles of Saturn's rings revolve around Saturn.'

Jeans goes on to describe the scene on a summer night in, say, the United States, with

> 'countless moonlets stretching as a luminous arch across the southern sky, whilst the shadow cast by the Earth upon it would produce a great 'bite' out of the luminous band. . . . By day this lunar belt would more or less hide the Sun for days or weeks, whilst in the winter only a faintly lit or perhaps scarcely perceptible belt would be visible. The reason for this is that the Sun would then be shining upon the other side of the innumerable particles, and only such light as penetrated would be visible together with such light as was reflected from the radiant Earth. This is, of course, supposing that there was any sunlight left, or the Earth anything better than a dark frigid waste at that remote epoch.'

It sounds dismal, but I am delighted to be able to tell you that it will not happen. As Jeans calculated, quite correctly, the process would take about 5,000 million years. He and all other astronomers of the 1930s believed that the stars would go on shining for millions of millions of years—but this is not true. Within 5,000 million years from now the Sun will have swelled out to become a red giant star, and both the Earth and the Moon will have been destroyed.

So we will never have Saturn-like rings. The time-scale is all wrong, and this shows just how much we have learned over the past half-century.

46 The Atmosphere of Ceres

Ceres, first and much the largest member of the asteroid swarm, was discovered by the Italian astronomer Piazzi on 1 January 1801. Piazzi was not hunting for a new planet; he was compiling a star catalogue, and it was only when he realized that one of his 'stars' was moving slowly from night to night that he alerted other observers. Ironically, a team led by Johann Schröter, in Germany, had already started a systematic search for a planet moving in an orbit between those of Mars and Jupiter. Ceres, of course, does lie in this part of the Solar System, and before long three more asteroids were found: Pallas, Juno and Vesta. The next discovery (that of Astræa) was delayed until 1845, but by now thousands of asteroids are known.

Most of them are very small. Ceres, with its diameter of 584 miles, has one-third the mass of the whole swarm. This is still too small for it to be classed as a bona-fide planet, but could it retain any sort of atmosphere?

Early observations indicated that it might. Schröter, who was an expert observer, found that it had a 'fuzzy' edge, and from this he inferred that there might be an atmosphere up to 500 miles deep; the same was true of Pallas, but not of Juno or Vesta. Then came observations by Admiral W. H. Smyth, a celebrated astronomer of mid-Victorian times, who wrote in 1844 that Ceres 'when observed with a telescope, appears something like a ruddy star of the eighth magnitude; raising the magnifying powers, under favourable circumstances, makes the disk—or rather its breadth of surface—perceptible, and leaves the inference of an extended atmosphere'.

Meanwhile Sir William Herschel, discoverer of the planet Uranus, had been looking at Ceres with one of his main telescopes, a reflector with a focal length of 20 feet. He too described the asteroid as reddish, but he was concerned mainly with the possibility of finding any satellites. Initially he believed that he had detected two.

A satellite of an asteroid would indeed be interesting, but we are now certain that Herschel had made one of his rare mistakes; his Cerean 'satellites' were simply faint stars in the background. We can also discount an atmosphere of the type suggested by Schröter and Smyth. This is because we have been able to find out the density of Ceres, because of its effects upon other members of the asteroid swarm, and it turns out to be only 2.7 times that of water. This means that the escape velocity is so low that there could not possibly be any dense atmosphere—and for that matter, Ceres pulls much too feebly to hold on to any satellite.

We cannot see any surface detail on Ceres, because the apparent disk is much too small, but we can analyse its light by means of a spectroscope, and some interesting facts have emerged. Though the surface is darkish, reflecting only 6 per cent of the sunlight which falls upon it, the American astronomer L. Lebofsky has found indications of a layer of frost, presumably ordinary water frost. Moreover, M. A'Hearn has now managed to detect a thin cloud of what are called OH ions around Ceres. This means that there is enough water ice on the surface to produce a very thin gaseous surround when it is warmed. But the density is so low that it corresponds to what we would normally call a good laboratory vacuum, and it is certainly wrong to say that Ceres has an atmosphere of anything like terrestrial type.

All the same, the discovery is worth noting, because it seems to reinforce the idea of a close link between larger asteroids, small Earth-grazing asteroids, and comets which have lost all their dust and gas. Just where Ceres fits into the overall picture is not yet clear.

Ceres is much the largest member of the asterid swarm. Here its size is shown compared with the British Isles.

What we need, of course, is a close-range picture obtained from a space-craft. No doubt this will be possible in the near future, and Ceres, the colossus of the minor planet family, is a prime target. Piazzi discovered it on the first night of the eighteenth century; it would be good to know that by the first night of the new century—1 January 2001—we may at last know what its surface is like.

☾47 A Very Old 'New Star'

Bright novæ, or new stars, have not been common recently. The last one flared up in Cygnus as long ago as 1976, though there have been many telescopic novæ since then, and even a few which have reached the limit of naked-eye visibility. Of course, a 'new' star is not really new at all: what happens is that the dim white dwarf component of a binary system suffers a tremendous outburst, and rises to many times its normal brilliancy before fading back to obscurity. A few novæ have become really spectacular; for instance, in 1918 Nova Aquilæ briefly outshone every star in the sky apart from Sirius.

Novæ have been recorded over the ages, and the remnants of almost all the 'modern' ones can be traced, though only with powerful telescopes. One very interesting case is that of the star of 1670, known as Anthelm's Star because it was first seen by the French monk of that name. It was in the little constellation of Vulpecula, the Fox, not far from Albireo in Cygnus, and it reached magnitude 2½, so that it was a prominent naked-eye object. It then disappeared, and was not traced again until 1981, when Canadian astronomers found wisps of nebulosity and a very faint star in the right position. There is no doubt that we are seeing what is left of Anthelm's Star, or, to give it its official designation, CK Vulpeculæ; but it is a very unusual remnant indeed.

It seems to be between 1300 and 2300 light-years away, and is just behind a cloud of interstellar dust which dims it considerably. The most remarkable characteristic is that it is only about 1/100 as luminous as our Sun, whereas all other old novæ are about equal to the Sun in output.

CK Vulpeculæ is therefore much the faintest nova remnant ever discovered. Apparently it is now made up of a dim, cool, red dwarf together with an even smaller, fainter, white companion. The two are moving round their common centre of gravity in a period of only about 3½ hours. In most such pairs there is a stream of gas from the larger, less massive member of the system through to the denser component, but in CK Vulpeculæ this 'mass transfer' has to all intents and purposes stopped.

One important consequence of this new work is that if CK Vulpeculæ is typical, then feeble old novæ may be much more common than has been thought. They could very easily be overlooked, and it follows that in this case actual nova outbursts may also be more frequent. So CK Vulpeculæ has taught us something which old Anthelm could never have imagined when he looked up at the Fox in 1670 and saw an unexpected addition to the sky.

C48 The Mousetrap that Squeaked

Space research can provide moments of hilarity. Hoaxes are only occasionally amusing, but I always liked the episode which occurred shortly after the launch of Russia's first artificial satellite on 4 October 1957, when an agitated American householder rang the Pentagon to say that a sputnik had landed in a tall tree in her garden. The report was taken seriously, and security officials were sent to investigate. The object turned out to be a large balloon, with 'upski' printed on the top and 'downski' on the bottom.

Sputnik 2 followed, and then came news of what seemed to be a third launch. It was claimed that the Finnish State Radio had picked up signals; London papers stated that a BUP correspondent in Helsinki had heard them, describing them as being similar to those of Sputnik 1, while extra confirmation came from West Germany and from a Norwegian coastal station at Tjömö. The London *Evening News* for 11 January 1958 carried a large headline: 'Sputnik III riddle. Fresh signals are picked up. Calls heard—then fade.'

It was left to the BBC's listening post at Tatsfield to solve the mystery. Officials there first said that they could not identify the signals, but believed the culprit to be an idling teleprinter. The bubble was finally pricked by the Swedish radio, which found that the teleprinter had become tired of sending 'Bleep! bleep!' and had started transmitting messages in ordinary Russian.

Now let us come to 2 April 1988 (not April the First, please note), when a Pan-Am jet-liner flying across the coast of Wales picked up what he thought was an SOS signal. This caused considerable surprise, and a message was sent through to Moscow. However, the Russian spokesman said that that

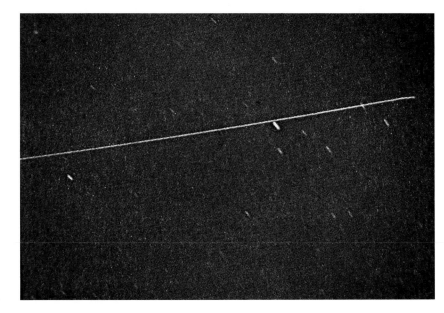

Trail of Sputnik 1. As the carrier rocket passed over Selsey in Sussex, a photograph of it was taken by Henry Brinton.

particular satellite should not be transmitting at all; it was not due to do so, and there was no conceivable reason why it should.

Moscow was baffled. So was everyone else, and so a helicopter was dispatched to see whether any alien interference could be tracked down. The results were positive, and within three hours the helicopter had traced a significant signal coming from a location near the town of St Asaph.

Enter that pillar of the Establishment, the country policeman, in this case PC Jones (are all Welsh policemen called Jones?). Somehow or other he 'homed in', and arrived at the house of a Mrs Mathers, whose previous connection with space research had been precisely nil. All became crystal clear. Mrs Mathers disliked mice, and had installed an electronic mouse-scarer which had been powerful enough to trigger off signals from the orbiting satellite. . . .

'It wasn't anything I'd expected', was her apt comment. 'After all, those shouldn't have been heard by anyone except mice and rats!'

☾49 Unscientific Satellites

Many artificial satellites have been sent up since the start of the Space Age, in 1957; some have been military, others scientific. There was a new departure in the late 1980s, however, with proposals which would, if implemented, have caused a great deal of damage.

The first of these was French. The Eiffel Tower celebrated its centenary in 1989, and the publicity company dealing with it (the Société Nouvelle d'Exploitation de la Tour Eiffel) planned what was to be called Project Light Ring, scheduled for launch by an Ariane rocket, from the French base at Kourou in South America, to a height of 500 miles. It would orbit the Earth once in 90 minutes, and would be highly spectacular. It was to be made in the form of a necklace, with a hundred 20-foot Mylar balloons connected by hollow tubes. The end product was to be a circular structure five miles across, so that it would have an apparent diameter of half a degree (equal to that of the Moon) and would look like a ring of one hundred stars of the first magnitude. It would last for three years, and would be visible for ten minutes out of every 90 minutes for two orbits after sunset and two orbits before sunrise. In fact, the whole of a summer night would be affected.

If the Light Ring proved to be impracticable, then the Eiffel Tower Company had an alternative—a space-art project, ARSAT, which was simply a curved reflecting sail launched into an 11-hour orbit. The curve would focus the Sun into a rotating X with a bright centre, projected on to the Earth over a region of from one to two thousand miles wide, drifting across a continent in three-quarters of an hour. The magnitude would be -15, much brighter than that of the full moon.

Next came the Celestis Corporation of Florida, whose organizers wanted to introduce a space mausoleum—so that the cremated remains of up to 15,000 people could be launched into satellites with polar orbits, and mourners could see their dead relatives 'orbiting for eternity', as the preliminary brochure put it.

Kourou, the European launching ground in French Guyana. It is near Devil's Island and has much the same climate!

Quite apart from the moral considerations, all these projects were serious threats to astronomy. There would be a near-certainty that one of the brilliant objects would come into the view of a telescope which was using very sensitive electronic equipment, and the immediate result would be to destroy the equipment completely. There could be no safeguard, and deep-space astronomical research would come to a stop.

It was difficult to see which of the two main proposals was the worst. French conceit seemed to transcend all bounds, though for sheer bad taste the

American Celestis Corporation could claim pride of place. Fortunately there was a storm of protest from scientists all over the world, and moreover neither of the companies could afford special rockets, so that to launch their schemes would have meant co-operation from the space authorities—which would certainly not be forthcoming. Finally the Eiffel Tower Company issued a statement to the effect that 'due to international pressure' the Light Ring had been abandoned, while for the moment nothing more has been heard from the Celestis Corporation in Florida.

Yet the threat of something of the sort in the future remains, and a careful watch will have to be kept in order to nip any harmful schemes in the bud. Moreover, one such satellite would open the way to 'commercials', and from this it would be only one step toward an illuminated procession of satellites streaking across the night sky spelling out the name of, say, a well-known brand of detergent!

Sirius—from Close Range

If I had to nominate the most successful space-probe of the twentieth century—so far—I think that my choice would be Voyager 2. It has, after all, had a truly magnificent career. It flew past Jupiter in 1979, Saturn in 1981, Uranus in 1986 and Neptune in 1989, sending back superb pictures and data from each; it achieved more than its makers dared to hope when it was launched, and we are still in touch with it. We hope not to lose contact until around the year 2020, though admittedly this may be rather over-optimistic.

At least we are certain of one thing: in the foreseeable future we will lose it. It is travelling so fast that it can never return to the Sun, and until it is either hit by a wandering object or else taken in tow by some alien civilization it will go on moving between the stars. It carries a record which will, we hope, tell any 'other men' its place of origin, but at best this cannot happen for a very long time.

In about 40,000 years' time, Voyager 2 will pass reasonably near a dim red dwarf star, but there is certainly no fear of a collision. Then, in about 290,000 years from now, will come the closest approach to Sirius, the brilliant Dog Star, which is 8.6 light-years from the Sun. This may seem a long time—but if Voyager 2 were travelling at 55 m.p.h., the legal limit for a car in the United States, the journey to Sirius would take not 290,000 years, but almost 180 *million*.

To us, Sirius shines as the brightest star in the sky (not counting the Sun, of course), but it is not exceptionally luminous; it has a mere 26 times the power of the Sun, and astronomers officially class it as a dwarf. Appearances can often be deceptive. Rigel, in Orion, does not look nearly so impressive as Sirius, but it could match at least 70,000 Suns. If Sirius were as remote as Rigel—around 900 light-years—it would be well below naked-eye visibility. Remember, we are seeing Rigel not as it is today, but as it used to be at the time of William the Conqueror.

Sirius. This photograph shows three views of Sirius. The bright star is overexposed (the spikes are, of course, photographic effects) but the companion—the Pup—is shown below the bright star. The Pup has the same mass as our Sun, but only 1/10,000 the luminosity of Sirius itself. It was the first known White Dwarf star.

Dwarf though it may be, Sirius has a diameter almost twice that of the Sun. It is also very hot, with a surface temperature of the order of 10,000 degrees Centigrade, and it is pure white, though as seen from Earth it often seems to flash various colours of the rainbow—simply because of the effects of our turbulent atmosphere, which, so to speak, 'shakes the starlight about', and causes twinkling.

Sirius is not a solitary traveller in space. It has a companion, only 1/10,000 as bright, but as massive as the Sun even though its diameter is less than that of a planet such as Uranus or Neptune. Because Sirius is called the Dog Star, the Companion is often referred to as the Pup—but it is a very substantial Pup, with a density over 100,000 times greater than that of water. It is a white dwarf star, which has used up all its store of nuclear energy and has no reserves left.

What would an observer on Voyager 2 see as he approached Sirius? Over thousands of years, the Dog Star would become brighter—and brighter—and brighter; of course it would not twinkle. Even from 4 light-years, the minimum distance from Sirius, the Pup would still be dim, but Sirius itself would shine almost as brilliantly as Venus does to us.

That would be all. Voyager 2 will go no closer to Sirius; it will start to draw away again, and once more the Dog Star will fade. In 800,000 years it will be no brighter than it seems from Earth, and eventually it will become so faint that an observer whose eyes are no better than ours will no longer be able to see it without optical aid.

Of course, this is all very fanciful. Just how Voyager 2 will move cannot be regarded as certain—how many wandering bodies are there in interstellar space?—and there is always the chance of a fatal collision. But it is quite possible that in less than 300,000 years from now a space-craft launched from Earth will be within striking distance of Sirius. It is a sobering thought.

☾51 Undiscovered Planets

The Sun's planetary family has nine known members, one of which is the Earth. Five have been known since antiquity; of the rest, Uranus was discovered in 1781, Neptune in 1846 and Pluto in 1930. There are also thousands of asteroids, but of these only one (Ceres) is as much as 500 miles in diameter, and cannot be regarded as being in the same class as a bona-fide planet. Even the status of Pluto is in doubt, because of its small size.

Can there be any more planets which we have yet to locate? All sorts of possibilities have been suggested, but all in all there seem to be only three sites where an extra planet might be lurking:

(1) Inside the orbit of Mercury, so close-in that it is drowned in the glare of the Sun.
(2) On the far side of the Sun, so that it always lies behind the Sun, as observed from the Earth, and cannot be detected.
(3) Beyond the orbits of Neptune and Pluto.

Little more than a century ago it was widely believed that an intra-Mercurial planet existed, and it was even given a name: Vulcan. The main Vulcan supporter was the French astronomer Urbain Le Verrier, whose views carried a great deal of weight; it was his calculations concerning the irregularities in the motion of Uranus which had led to the discovery of Neptune by Johann Galle and Heinrich D'Arrest, at the Berlin Observatory. Flushed with this triumph, Le Verrier turned his attention to the movements of Mercury. He found irregularities there too, and attributed them to the gravitational pull of an inner planet which moved round the Sun at a distance of about 13,000,000 miles in a period of 18 days.

A planet so close to the Sun would be observable only if it passed in transit across the solar disk, as both Mercury and Venus can sometimes do. On 26 March 1859 a French amateur named Lescarbault claimed that he had actually watched such a transit. Le Verrier went to see him at his home in Orgères, and was impressed, even though Lescarbault hardly qualified as a professional scientist; he doubled as the local carpenter, and used to record his observations on planks of wood, planing them off when he had no further use for them!

Vulcan! In 1878 there was a total eclipse of the Sun. While the Sun was blotted out, L. Swift sketched the surrounding area, and believed that he had located Vulcan. In fact, he had simply seen faint stars.

Le Verrier was convinced in the real existance of Vulcan, but it has never been seen again, and there seems no doubt that what Lescarbault saw (if anything) was an ordinary sunspot. Moreover, the irregularities in the movements of Mercury have been explained satisfactorily by Einstein's theory of relativity. So Vulcan has been regulated to the status of a ghost; the only bodies to move in those torrid regions are occasional comets and asteroids, which do not stay there for long.

The idea of a counter-Earth, staying exactly behind the Sun, is very old; it goes back to Greek times. However, the alignment would not last for long, and would soon be destroyed by the gravitational pulls of Venus and the other planets, so that counter-Earth would come back into view. This leaves us with our last potential site, the cold wastes of the Solar System out beyond Neptune and Pluto.

Pluto has proved to be so small and lightweight that it could not possibly have measurable effects upon the motions of giant planets such as Uranus and Neptune. The fact that it turned up in much the expected position must therefore be purely fortuitous—unless, of course, there really is another planet at a still greater distance. Many astronomers believe that there is; we usually refer to it as Planet X, and periodical searches for it are made. For example, J. Anderson, in the United States, has concluded that Planet X is about five times as massive as the Earth; it moves round the Sun in a very eccentric orbit, taking about 1000 years to complete one revolution.

New calculations have now been made by a Russian team led by Vladimir Radziyevski. Instead of concentrating on the movements of known planets, the Russians have worked upon the movements of comets, which are very flimsy and insubstantial, so that they can easily be 'pulled around'; most of them have very eccentric orbits, and most bright comets spent much of their time well beyond the main planetary system.

Radziyevski has concluded that there is not one outer planet, but two. He calls them X1 and X2. Each, he believes, is about fifty times as massive as the Earth (making them more massive than Uranus or Neptune, and roughly half as massive as Saturn), with periods of around 2000 years. His Planet X2 could be the same as Anderson's single Planet X. Both have paths which are almost perpendicular to those of the Earth and the other planets of the Sun's family.

But do any of these hypothetical bodies really exist? And what about the luridly named Nemesis, which is said to have at least 1000 times the Earth's mass, so that it qualifies as a dark star and is claimed to be heading toward us? Nemesis, too, depends upon the movements of comets, but I must admit that I am a complete sceptic. In my view Planet X probably does exist, and I have an open mind about two, but I am no believer in the menacing dark star.

52 New Thoughts on a Remarkable 'Kid'

In my first book of random essays I devoted a short section to Epsilon Aurigæ, an ordinary-looking star lying not far from the brilliant Capella, and well seen from Britain for a large part of the year; during winter evenings it is not far from the zenith or overhead point. It is one of three stars which make up a small triangle, and are known collectively as the 'Kids'. (Epsilon Aurigæ does actually have an old Arabic name—Almaaz—but it is hardly ever used.) I do not apologize for saying more about it now, because we have had some fresh thoughts about it.

Normally it shines as a star of the third magnitude, very slightly brighter than the second Kid, Eta Aurigæ. However, it is not constant. Every 27 years it fades slowly by almost a magnitude, remaining at minimum for a year before recovering. The latest minimum began in July 1982, and did not end until June 1984.

Epsilon itself is a yellowish supergiant, at least 60,000 times as luminous as the Sun, and about 3200 light-years away. Associated with it is an invisible companion, and it is this companion which passes in front of the supergiant every 27 years, cutting off part of its light; we are dealing not with a true variable star, but with an eclipsing binary. But what is the nature of the companion? It is completely invisible, and does not radiate even at infra-red wavelengths, but we can estimate its mass, which proves to be about 20 times that of the Sun (as against 35 Suns for the supergiant). Few stars are as heavyweight as this, and the secondary would certainly be visible if it were a normal star. We have a problem to solve, and there have been several plausible theories:

(1) The companion is a very young star which has not yet become hot enough to radiate perceptibly, and is still condensing. In this case the length of the eclipse would give a key to the diameter of the companion, which works out at 2,000 million miles, large enough to engulf the orbits of all the planets in the Solar System out as far as Saturn. It would be much the largest stellar body known. Yet even at mid-eclipse the supergiant can still be seen, and it follows that the companion would have to be completely transparent, which is not likely.

(2) The secondary body is a black hole, i.e. an old, collapsed star which is pulling so strongly that not even light can escape from it. But with a black hole, material which is about to be sucked in would be heated so strongly that it would emit high-energy radiation, and nothing of the sort has been found with Epsilon Aurigæ.

(3) The secondary is a disk-shaped cloud, so placed that when passing in front of the supergiant it bisects it, leaving some of it visible. This would be a strange coincidence, however, and moreover such a cloud would be unstable; we would also expect to see it in infra-red, which we do not.

(4) The companion is a smallish, hot star surrounded by a shell of gas, and it is this shell, so heated that it has become opaque, which causes the eclipses. The problem here is that we can hardly visualize a shell as large as the eclipsing companion would have to be.

The latest model is due to the work of the Italian astronomer Steno Ferluga, of Trieste, together with Sean Carroll of the United States and Edward Guinan of Spain. After analysing the observations of the 1982–4 eclipse they believe that Epsilon Aurigæ does not have a genuine companion, and that the eclipses are due to a thin, tilted disk with a partly transparent inner region and a clear zone in its centre. This is just what we would expect to find with a protoplanetary cloud—that is to say, material which is about to start condensing into planets. So could Epsilon Aurigæ be a candidate for a planetary system?

This is certainly speculative, and a star as luminous as our 'Kid' does not seem suitable to be the centre of such a system, but we cannot be sure. To quote Sean Carroll: 'Great strides toward understanding the nature of planet formation become possible with the discovery of an eclipsing binary with such an object.' In any case we must agree that Epsilon Aurigæ is unique; nothing else quite like it is known. With the naked eye, or with an ordinary telescope, it looks like an ordinary star, but it is anything but that.

☾53 The Legacy of Velikovsky

Science attracts eccentrics in much the way that an open jamjar will attract wasps. I hesitate to use the term 'crank', because most of them are sincere, harmless and amusing; I prefer 'independent thinker'. Occasionally, however, we find someone of unorthodox views who manages to exert real influence. In my experience there have been three: Hans Hörbiger, Trofim Lysenko and Immanuel Velikovsky.

Hörbiger, an Austrian engineer, produced his book *Glazialkosmogonie* in 1913. I cannot read German, so I have had to depend upon translations, but it appears to be long, tortuous and utterly without value. (My knowledge of the Teutonic language is limited to 'Damit', 'Besonders' and 'Donner und Blitzen'.) According to Hörbiger, the Sun is the most important body in the universe, and the stars are chunks of ice. Indeed practically everything, apart from the Sun and the Earth, is made up of ice; the Martian canals are simply cracks in the upper ice-sheet. The planets are having to push their way through rarefied hydrogen, and when a planet spirals down to hit the Sun it produces a sunspot. The Moon is itself spiralling toward us, and will eventually crash-land. More-over there have been at least six previous moons, all of which have suffered the same fate; each time there is an impact the result is a major extinction, such as that which wiped out the dinosaurs 65,000,000 years ago. Our present Moon was captured by the Earth about 13,000 years ago, attended by earthquakes and volcanic outbursts which effectively destroyed the civilization of Atlantis.

Hörbiger's WEL (Welt Eis Lehre, or Cosmic Ice Theory) appealed to the Nazis; after all, Adolf Hitler himself had a strong mystical streak. During the 1930s, indeed, WEL became so entrenched that the German Government actually had to issue a statement to the effect that one could still be a good Nazi without embracing Hörbiger's ideas. I never met Hörbiger, but I gather that it was difficult to argue with him, because anyone who disagreed with WEL was automatically classed as a mortal enemy. One of his followers was a German astronomer named Philipp Fauth, who produced a large map of the Moon and whose influence helped to keep the ice picture in the public eye.

Fauth died during the war, but WEL survived. As recently as 1953 it published a pamphlet claiming that 'the final proof of the whole cosmic ice theory will be obtained when the first landings on the ice-coated surface of the Moon take place'. I am not sure how the remaining Hörbiger supporters reacted when Neil Armstrong stepped out on to the Sea of Tranquility and found that he had no need to put on skates.

I will say little about Trofim Lysenko, because he was a geneticist, and my ignorance of genetics is complete—but it is quite true to say that Lysenko's theories, accepted by Stalin and regarded as authentic, effectively stopped the progress of genetic science in the Soviet Union for several decades, particularly as any opponent was removed to a labour camp and was unlikely to emerge. So let me pass on to Velikovsky, who was undoubtedly the most influential of all the 'independent thinkers'.

The First Men on the Moon: Neil Armstrong and Edwin Aldrin.

He was born at Vitebsk, in Russia, in 1895, and subsequently became a psychoanalyst, working for some time in the Middle East before emigrating to America and spending the rest of his life there. His first major book, *Worlds in Collision*, was published after the war by the Macmillan company, and was presented as cold, sober fact—to the extent that educational establishments threatened to boycott Macmillan until Velikovsky's magnus opus was removed from their list.

Velikovsky's basic outlook was religious, and depended upon the Bible. According to his theory, Venus was originally shot out from Jupiter, and was a comet rather than a planet; this happened only a few thousand years ago. At first the comet Venus had an elliptical path, and this led it into a series of close encounters. It passed near the Earth in 1500 BC, at the time when the Israelite Exodus was in full swing, and the result was that the spin of the Earth was slowed down, leaving the Red Sea high and dry for long enough to allow the Israelites to cross. Next came huge upheavals as the Earth squirmed under the gravitational pull of the comet Venus. Petrol rained down—indeed, the petrol we use today comes from this period—and there was general chaos.

Conveniently, the Earth's rotation started up again just in time to swallow up the Egyptians who were in hot pursuit of the retreating Israelites. But this was not all. Having made its presence felt, the comet Venus withdrew, but came back for a second visit two months later to produce thunder, lightning and other special effects so much in evidence during the first announcement of the Ten Commandments on Mount Sinai. Subsequently, some of the materials in the comet's tail rained down on Earth, producing manna upon which the Israelites fed for the next forty years.

One might have thought that this would be enough. But no! The comet Venus continued its erratic course, coming back to see us several times—once, for example, to shake down the walls of Jericho. However, the day of reckoning was at hand. The comet Venus collided with Mars, so that it stopped being a comet and started being a planet. Not to be outdone, Mars itself moved closer to the Earth, and nearly scored a bull's-eye in 687 BC. Further encounters took place, linked by Velikovsky with various episodes in the Old Testament; on one occasion the Earth turned head over heels, so that for a while the Sun rose in the west and set in the east. Finally, things settled down. Mars went back to its original path, presumably satisfied; Venus retreated to its present circular orbit, and all was calm once more.

What can one make of all this?

The immediate problem is that Velikovsky's ignorance of science is so complete that there is no basis for argument. His time-scale is ludicrous; he does not know the difference between a planet and a comet; his mathematical knowledge seems to be limited to addition and subtraction. What he could do, as he showed in *Worlds in Collision* and subsequent books, was to write plausibly and well. Unfortunately, the fact that his book had been issued by a reputable publisher as a proper scientific text caused some astronomers to over-react. Instead of ignoring Velikovsky, they attacked him, and there were even official conferences called to challenge his theories. This was a pity; it would have been far better to smile sadly and dismiss him as a totally well-meaning eccentric who was mentally several sandwiches short of a picnic.

The real damage is that proper 'catastrophic' theories, such as those of Victor Clube and Bill Napier, have been tarred with the same brush, at least in the popular mind. So even though Dr Velikovsky has now died, convinced to the last minute that he was the greatest thinker of all time, his influence lingers on.

C54 A Curious Stellar Family

Binary star systems are common in the Galaxy. Many of them are made up of two components, moving round their common centre of gravity much as the bells of a dumb-bell will do if you twist them by their joining bar. We also have triple, quadruple and even multiple systems. Recently there have been studies of a faint but very unusual 'stellar family' in the constellation of the Lion, known as XY Leonis.

It has been known for some time that XY Leonis is what is called a contact binary, made up of two stars which are actually touching each other as they whirl around. Tremendous activity is to be expected, but now Samuel Barden, of the Kitt Peak Observatory in Arizona, has come up with an unexpected solution, claiming that XY Leonis is four stars rather than two.

The main pair has a revolution period of 6 hours, which does to some extent account for the great activity. Both components are of spectral type K, rather cooler than the Sun and orange in colour. About as far from this pair as Saturn is from our Sun we find a smaller pair of M-type red dwarf stars, more than 1,400,000 miles apart, revolving round each other every 18 hours as they trace a 20-year-long orbit round the first pair.

This is indeed a weird arrangement, and it is interesting to examine the view which would be had by an observer living on a planet moving round the main pair. The 'sun' would be made up of two orange stars, spinning visibly and deformed tidally into a sort of figure-8. The senior member would have 9/10 the mass of our Sun and about one-third of its luminosity, while the companion would have less than half the Sun's mass and one-fifth of its luminosity. At equivalent distance, the combined brightness of the pair would be about half that of the Sun as seen from the Earth. The companion pair would appear as two very brilliant stars, the larger with half the mass and less than half the diameter of our Sun with a mere 2 per cent of its luminosity. It would look a thousand times fainter than the Sun as seen from Earth, but this would still be three hundred times brighter than our full moon. As Dr Barden commented, 'You wouldn't see many other stars when those were up at night'.

It is not easy to work out the history of XY Leonis, which is decidedly unusual. Whether there are any planets in the system we do not know, but at least we can be sure that the view from there would be bizarre by any standards.

C55 A Bolt from Midas?

Meteorites are unpredictable things. Some are made of iron, some of stone, while most are mixtures; probably they come from the zone of minor planets or asteroids, which lies in the wide gap between the orbits of Mars and Jupiter. Meteorites are not associated with shooting-star meteors, which are cometary débris and which burn out long before they hit the Earth's surface. The two classes of objects are quite distinct.

Large meteorites may produce craters, such as the famous structure in Arizona, once described by the Swedish scientist Svante Arrhenius as 'the most interesting place on Earth'. Most museums have meteorite collections, and before the start of the Space Age these missiles from afar represented the only non-terrestrial material which we could actually handle and analyse.

There is no authentic case of anyone having been killed or badly hurt by a tumbling meteorite, but houses have been hit now and then. The latest case was that of April 1990, in the town of Enschede in Holland. The missile, now known officially as the Glanerbrug Meteorite, smashed through the roof of the house. It was no more than about one kilogramme in mass, but

A little-known meteorite. This one fell long ago in China, and was discovered in 1976; it weighs 1770 kg, and is known as the Jiling Meteorite.

it seems to have been part of a 500-kilogramme object which had a diameter of about two feet. Chemical analysis showed it to be stony, and according to calculations it entered the Earth's upper air at a speed of at least 60,000 miles per hour.

How do we know this? Well, its fall was observed. As it flashed across the sky it was seen by many people, mainly members of the Dutch Meteor Society, and almost two hundred reports were received. Peter Jenninskens, of the Leyden Observatory, used these to construct an orbit—and to make a most interesting suggestion.

Jenninskens found that the orbital inclination, i.e. the tilt of the object's path relative to that of the Earth, was around 40 degrees. This is unusually high. If we assume (as most people do) that meteorites are essentially the same as asteroids, we may look for an asteroid which might have 'shed' a part of itself. The best candidate is a small body called Midas, which is one of the few asteroids to have an inclination of about this value, and which can make close approaches to the Earth.

Midas belongs to what is called the Apollo class of asteroids. These objects have orbits which cross that of the Earth, so that we are not entirely immune from collisions with them. Many of them are now known (Apollo, discovered in 1932, was the first; hence the name) and all are small; Midas, found by Charles Kowal in 1973, has a diameter of approximately one mile. Its path takes it from 58,000,000 to 166,000,000 miles from the Sun in a period of 2.4 years. Even when it passes close to us, as will next happen in 1992, it appears as no more than a starlike point, but its movements are well known, and we are easily able to keep track of it.

Obviously we cannot be sure that the Glanerbrug Meteorite is a broken-off piece of Midas, but the Dutch astronomers regard it as a distinct possibility. At any rate, we may be sure that the house in Enschede will be long remembered—and we must be glad that nobody happened to be in the line of fire when the missile made its dramatic entry through the roof!

C56 The Youngest Star

In astronomy we are used to reckoning in thousands of millions of years. The Earth, for example, is at least four and a half thousand million years old, while the universe in its present form dates back for at least 15,000 million years. Therefore it comes as a surprise to find that we have now identified a star—or, rather, a proto-star—with an age of only a few thousands of years. This makes it the youngest star known to us.

The story began in 1983, with the career of the Infra-Red Astronomical Satellite IRAS. Though it operated for less than a year, it sent back a vast amount of data, and ranks as one of the most successful space vehicles ever launched. Its detector studied cool material, not yet hot enough to radiate at optical wavelengths, and in a rather obscure nebula in the constellation of Perseus it found a discrete source which was listed as IRAS-4. Its distance was known—around 1100 light-years—because this is the distance of the nebula itself, which is known as NGC 1333. (NGC stands for New General Catalogue, though by now it is more than a century old.) But it is only in very recent times that the object has been re-examined.

To study it Colin Aspin, Goeran Sandell, Bill Duncan and Ian Robson used the JCMT or James Clerk Maxwell Telescope on the summit of Mauna Kea, in Hawaii, at an altitude of almost 14,000 feet. The JCMT operates at millimetre wavelengths, in between infra-red and the radio range, and it looks like a radio telescope; of course it does not produce a visible picture in the way that an optical telescope does. It is the world's most powerful instrument of its kind, and it has been a major success.

IRAS-4 proved to be of special interest. The team found that it is a star so young that it is still surrounded by a protective cocoon of dust and gas, so that it does not emit visible light; millimetre wavelengths can penetrate the cocoon, and so we can be sure that we really are dealing with a proto-star.

'Proto-star' means that the star is still in the process of formation. Material from the surrounding shell is falling on to it under the action of gravity; as

Observatory of the JCMT (James Clerk Maxwell Telescope) atop Mauna Kea in Hawaii. The membrane is in place; it is seldom removed even during observing.

Observatories on Mauna Kea. The James Clerk Maxwell observatory is in the foreground; to the left, the American microwave telescope; on the skyline the visual and infra-red instruments to the right, the Keck to the left.

The James Clerk Maxwell Telescope (JCMT) on Mauna Kea, 1991.

it hits the proto-star energy is released, and this is the source of IRAS-4's present warmth. The proto-star itself is not nearly hot enough for nuclear reactions to be triggered off. For this you need a temperature of the order of 10,000,000 degrees, so that IRAS-4 has a long way to go; it may be a million years before the cocoon is blown away, nuclear reactions start, and we are left with a normal star, considerably more massive than the Sun.

Using the JCMT at wavelengths of between 0.35 and 2 millimetres, the British team found indications of carbon monoxide and carbon sulphide, showing that there is an outflow of material from IRAS-4 into a cavity in the surrounding gas. This outflow may be driven by energy released in the disk of material which is moving near the surface of the proto-star. In the end, about 90 per cent of the material in the dust cocoon will be expelled by these outflows.

It is all very interesting, and shows how much we can learn from these great new telescopes. For the first time we have made direct observations of the birth of a star.

57 Blowing Up the Moon

Now and then a professional scientist comes up with a theory which seems to be drawn straight from the script-writers of the Goon Show. The tradition may be said to have started with Sir William Herschel, discoverer of the planet Uranus and arguably the greatest astronomical observer who has ever lived; he was firmly convinced that the Sun is inhabited, and that below the fiery surface there is a cool, pleasant region. Herschel died in 1822, but he has had a modern supporter in the Rev. P. H. Francis, of Sussex, who is a mathematician with an excellent Cambridge degree, but who has no patience with the idea that the Sun is hot. To quote his words from his book *The Temperate Sun*: 'The popular notion that the Sun is on fire is rubbish, and merely a hoary superstition, on a par with the belief in a flat earth . . . It rests on no sure basis of evidence, and if it is discarded, great simplifications become possible in the science of astronomy, geology and physics.' He also pointed out that there is a good analogy with an electrical generation station. When you switch on your electric fire, the bars become hot and glow, but the actual power comes from the generation station—and the station itself certainly isn't hot.

The latest really bizarre theory, produced in 1990, comes from a Professor Alexander Abian of Iowa University in America. He too is a mathematician, and he is not impressed with the Earth's climate. His solution is quite simple: blow up the Moon!

I quote Professor Abian:

'We can then shift the Earth into a more desirable orbit. We could stop brutal winters and scorching desert summers simply by blowing up the Moon. We have been held hostage in the same orbit for five billion years, but there is no reason to believe that it is an optimal one so far as the ecology of our planet is concerned. Blowing up the Moon would alter the gravitational parameters and create a less extreme climate.'

Of course, there is no need to destroy the Moon completely; you might break off part of it and dump it in the Pacific Ocean, or even split the Moon in half. Professor Abian again:

'We would only need to destroy large chunks of it, split it in two and land some of it in the Pacific, preferably gently.'

I can give you no details about the professor himself, except that he was born in 1927 and is of Armenian origin. As soon as his statement was issued—apparently it was submitted to the United States Government!—I was questioned about it. I pointed out, gently, that even if the Moon were removed it would not alter the tilt of the Earth's axis in the way that the professor seems to believe. Moreover, the energy needed to destroy the Moon would certainly destroy the Earth as well, even if we had the faintest idea of how to do it. The British Meteorological Office commented that a moonless Earth would be 'bleak and tideless', and a spokesman for the British Association for the Advancement of Science, struggling nobly to keep a straight face, asked what would happen if the experiment went wrong. Predictably, Professor Abian was unrepentant. 'People don't seem prepared to sacrifice the Moon for a better climate. It is inevitable that the genius of man will one day accept my ideas.'

Somehow I doubt it. I was asked by a radio producer to sum up the whole scheme, and I could only say that in my view it had to be described, together with its proposer, as NAFC. In case you don't know what that means, NAFC stands for 'Nutty As a Fruit-Cake'.

C58 A Tale of Two Fires

High on Palomar Mountain in California, not too far from the city of San Diego (not far enough, the dark-sky conscious astronomers claim!) is one of the world's major observatories. Here we find the world's largest really good single-mirror telescope, the 200-inch, as well as a host of other instruments.

To go there you have a long drive up a steep but pleasant mountain road. There is not a great deal of vegetation near the top, but there is enough to catch fire—and in the summer of 1989 this is what happened. Near Vail Lake, on the mountain slopes, lies the Dripping Springs camp-ground. It is popular with visitors, and it seems that around 11.30 in the morning of 29 July someone having sharp-shooting practice set off a spark which began a serious fire.

It was detected before long, but strong, erratic winds made it go out of control. There is a great deal of scrubby material, plus very old, large trees called Douglas firs; there is also the 400-acre Agua Tibia Research Natural

Café on Palomar, once run by the flying saucer king George Adamski. I took this picture in 1975—it has changed hands since then.

Area in the Eagle Crag region, a federal ecological preserve dedicated to bald eagles and spotted owls, both of which qualify as endangered species. There were some dwelling-houses. And there was the Observatory. . . .

Roads were closed, and firemen arrived. In the end there were 2800 of them, drawn from all over the area. In spite of their efforts the fire spread, and by 6 August it was within two and a half miles of the observatory. I asked Dr Eleanor Helin, who was there to carry on with her observations of near-Earth asteroids, whether the danger was real. She said: 'It sure was. For a while anything might have happened.'

Luckily, Nature came to the rescue. Richard Markley, the US Forest Service fire officer, said that by pure chance 'the winds advanced the fire to the east, instead of south toward the direction of the Observatory', but it was a very close call indeed.

Gradually the fire-fighters started to win the battle. On 7 August, Markley said that 'the storms could be accompanied by more winds and lightning that could set off embers, but we're hopeful that they will bring in rainfall to dampen the ground'. This is what happened, and the threat to the Observatory was regarded as over by 9 August. Two days later the fire was out, though teams stood by to make sure that there were no unexpected flare-ups.

All in all, 15,643 acres were destroyed, plus three houses. In the 90-degree summer heat it was said that the landscape looked like the bottom of a barbecue pit—charcoal-coloured, burned to a crisp and partly covered with white ash. The short scrub bush which survived was orange-red from the material sprayed down on to it from the fourteen air-tankers. On the credit side, it was a low-intensity burn which left most of the firs intact, and it did not do lasting damage to the nature reserve—though ecologists were quick to come in to see which areas needed re-seeding, and to stop any rain erosion of the gulleys which had been dug to stop the spread of the fire. The total cost was at least 8,000,000 dollars.

The General Secretary, D. McNally, addressing the emergency meeting at La Plaza. Right: The President, Y. Kozai, Past President, J. Sahade and Chairman of the LOC, R. Méndez.

Palomar: the 200-in reflector, as I photographed it in 1986.

But the Observatory was safe. The Palomar Mountain General Store, nestling on the slopes and once the haunt of flying saucer king George Adamski, cashed in with T-shirts, each of which carried a picture of the Observatory with the design: 'FIRE OF '89. JUST WHEN YOU THOUGHT IT WAS SAFE TO GO BACK—PALOMAR II, THE SEQUEL. FIRE OF '89. BYOM'. Most people had forgotten the earlier fire of October 1987, which had blackened over 16,100 acres. And in case you do not understand the meaning of the final BYOM, it stands for 'Bring Your Own Marshmallows'. Please do not ask me why!

Another astronomical fire was that of 31 July 1991. It was very minor compared with the Palomar episode, but I cannot resist saying something about it, because I was involved myself.

The controlling body of world astronomy is the International Astronomical Union (IAU). I have been a member since 1966; there are General Assemblies every three years, held in different countries. The 20th General Assembly, that of 1988, was held in Baltimore (at that time we were all looking forward to the launch of the Hubble Space Telescope, and the main Hubble headquarters is there). In 1991 it was the turn of Buenos Aires, in Argentina—the first time that an Assembly had taken place in South America. One of the regular features is a daily newspaper, in which results are summarized, comments made, and general announcements and decisions given out. It is distributed every morning to each delegate. There were around 1500 delegates at Buenos Aires, and I had been asked to edit the Newspaper. With my colleague John Mason I arrived days early, and made arrangements.

Our editorial office was at the San Martín Centre, where the main meetings were held. The Assembly was due to end on 1 August, with the passing of the final Resolutions, which had to be printed in our final issue. Naturally, they were not ready on the previous evening. Therefore we decided that we would plan, key-in and design most of the last issue, and add the Resolutions as soon as we could lay our hands on them in the following morning. We finished our labours at 2 a.m., and then, pausing only for a quick glass of the excellent Argentinian wine, returned to our hotel.

We had planned to be back at the Centre by 8 a.m., and, as usual, we walked there (it took only a few minutes). As we came down Sarmiento Street, we saw clouds of smoke billowing up ahead of us. 'I hope that isn't the Centre on fire', one of us said jokingly. But it was—and as we arrived we saw crowds of people milling around, fire-engines pumping water into the basement, and a general scene of utter chaos. There were even firemen scaling up and down ladders with cat-like agility.

Our first instinct was to find out whether anyone had been trapped when the fire started. Mercifully few people had been inside, and there were no casualties (though later we heard that two firemen had been rendered unconscious by smoke). Our second thought was: 'What about Issue No. 10?' We had all the first part ready for printing; all we needed was to salvage it from the second floor of the Centre. The first attempt was abortive; there was too much toxic smoke. The second attempt succeeded, and by noon we were standing forlornly on the soaking steps of the Centre, with two

Fire over the IAU! The scene as we rescued our one little computer and assembled outside the smouldering Conference Centre!

computers, a mass of sodden papers and miscellaneous pieces of equipment. We carried everything back to our hotel, and then set to work to finish the issue. The Resolutions, of course, could not be included, because the main IAU computer was still in the smoke-filled centre.

One conversation stands out in my mind; it was between us and the representatives of the Local Organizing Committee. It went as follows.

> PM. Where will the final Assembly be held, and when?
> LOC. It will have to be in the Plaza [the building opposite] at 10 a.m. tomorrow. Can you put this into the newspaper?
> PM. Yes. Where will they have to collect their newspapers, so that they can find out the time and place?
> LOC. Oh—it will have to be 10 a.m., at the Plaza.

It was a perfect example of circular reasoning. In the event we distributed our last issues at various hotels, and I think that everyone received one, but it was an eventful finale—never again can it be suggested that an IAU General Assembly is dull!

☾59 The Star in the Indian Bowl

If you had looked into the sky during July 1054 you would have seen something very unusual—a star so bright that it remained visible even in the middle of the day. We know what it was: a supernova, representing the death-throes of a very massive star which had blown up, and sent most of its material hurtling away into space. The 1054 supernova was recorded in China, though apparently not in Europe. Gradually it faded, and once it dropped below naked-eye visibility it was lost (there were no telescopes in those days!) but we see its remnant today in the form of the gas-cloud of the Crab Nebula, in Taurus the Bull. Right inside the gas-cloud is a neutron star or pulsar, the core of the old, exploded star, which is spinning round rapidly and sending out radio waves.

The Crab Nebula, the remnant of the supernova of 1054. It contains a pulsar and is 6,000 light years distant.

We would like to have more records of the supernova itself, and we may now have found an unexpected observation, coming from New Mexico in the United States. Over half a century ago an Indian burial bowl was discovered there, seven inches in diameter, and apparently made between the years 1000 and 1070. Archæologists from the University of Texas at Austin have been

The White Spot on Saturn. On 9 November 1990 this picture was taken with the Planetary Camera of the Hubble Space Telescope. By then the spot had spread along the equator, and had become a bright band rather than a well-defined spot. Reproduced by kind permission of ESA and NASA.

Saturn, as seen by the Hubble Space Telescope—a better view than can be obtained from the ground. Reproduced by kind permission of ESA and NASA.

Deployment of the Hubble Space Telescope, on 25
April 1990, from the Space Shuttle Discovery. It was
placed into an orbit 380 miles above the Earth.
Reproduced by kind permission of NASA.

Global view of the surface of Venus, centred at longitude 270° E. Magellan synthetic aperture radar mosaics from the first cycle of Magellan mapping are mapped on to a computer-simulated globe to create this image. Simulated colour, based on colour images from the Russian Veneras 14 and 15, is used to enhance small-scale structure. Reproduced by kind permission of NASA.

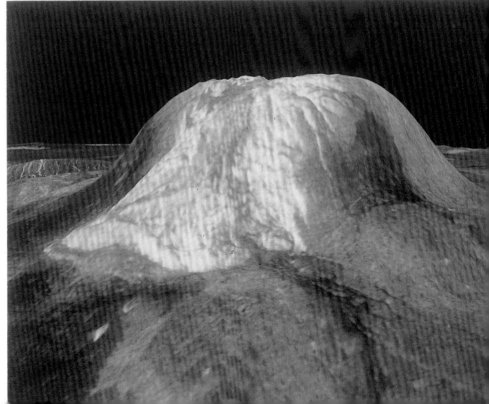

Computer-simulated view of Gula Mons, on Venus. The viewpoint is located 68 miles S.W. of Gula Mons at the same elevation as the summit, 1.9 miles above Eistla Regio. Lava flows extend for many miles across the fractured plains. Gula Mons, a 1.9-mile high volcano, lies at 22° N, 359° E, in western Eistla Regio. Magellan synthetic aperture radar is combined with radar altimetry to produce a three-dimensional map. Reproduced by kind permission of NASA.

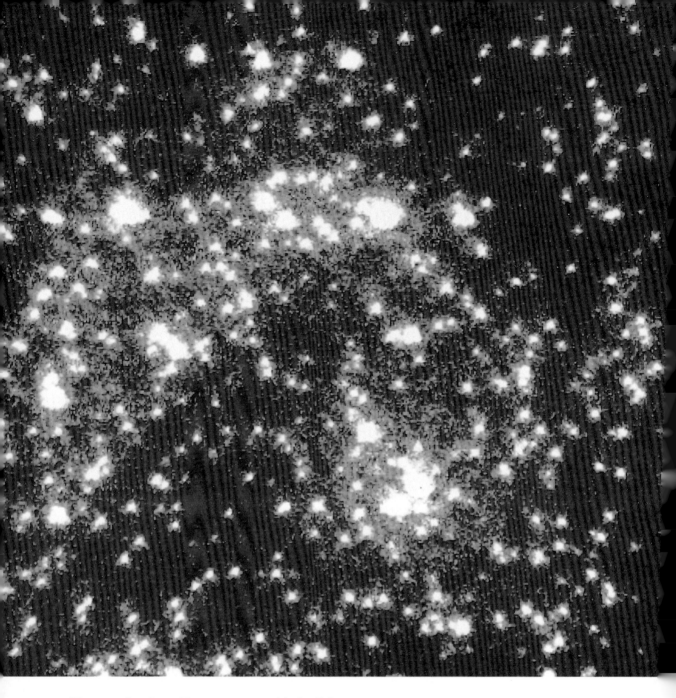

*Blue stragglers in 47 Tucanœ, as seen with the Faint
Object Camera on the Hubble Space Telescope.
Reproduced by kind permission of ESA and NASA.*

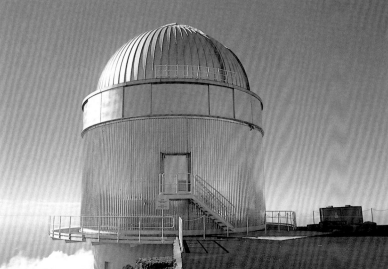

▲

Star trails. I took this picture from La Silla, in Chile—one of the domes is shown—and so Orion is 'upside down' to the British viewer. Orion is to the upper left of the dome. The very bright trail to the left is Jupiter.

▶

Dome of the Nordic Telescope at La Palma, as I photographed it in 1991.

Beta Pictoris. This is an artist's impression of the near-stellar region. The reddish outer ring is a diffuse gas disk in a stable orbit; this surrounds an inner disk which is slowly spiralling down toward the star. The white comet-like features in this bluish disk are dense streams of gas spiralling downward. The outer filamentary structures may be either an expanding gaseous halo, or else more local foreground features. Reproduced by kind permission of ESA and NASA.

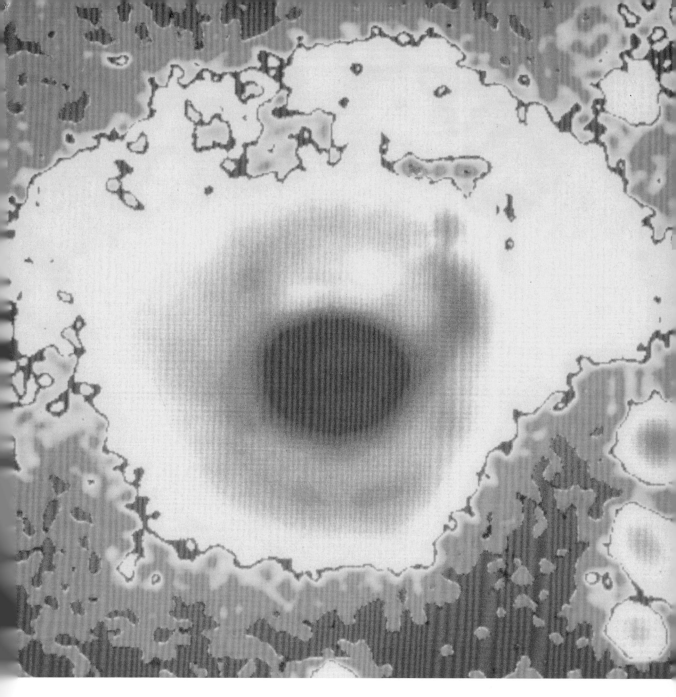

The Antique Nebula. This nebula surrounds the supernova 1987a in the Large Cloud of Magellan. Formerly, this star was a cool red supergiant; it shed its outer layers, and became a hot, blue supergiant which exploded. This picture shows the structure as it was before the outburst. In a few years the expanding envelope of the supernova will overtake this 'antique' nebula and destroy it.

The picture is a false-colour image, obtained with the NTT telescope at La Silla in Chile; it is doubtful whether any other current telescope could reveal as much detail. Reproduced by kind permission of the European Southern Observatory.

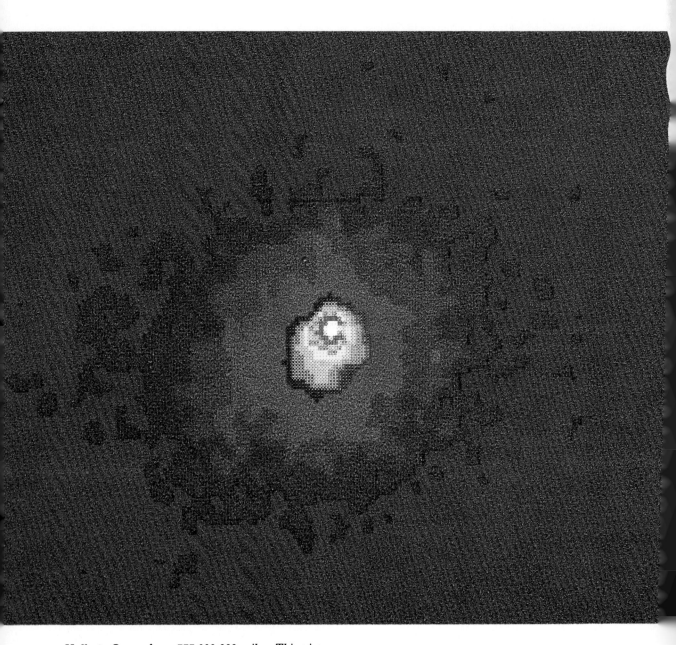

Halley's Comet from 777,000,000 miles. This picture was taken with the Danish 60 in telescope at La Silla, in Chile. It is of course a false colour picture, obtained by combining 50 CCD frames spread over 19 nights in April and May 1988—before the mysterious outburst! At the centre, the nucleus (white) is within the inner coma (yellow) which is in turn surrounded by the outer coma (green). The magnitude of the nucleus at the time was 23. Reproduced by kind permission of the European Southern Observatory.

looking at it, and have claimed that on it there is a representation of the supernova of 1054.

The bowl shows the stylized form of a rabbit—and it was a rabbit which most American Indians used to symbolize the Moon, because they felt that this was the shape indicated by the light and dark patches on the disk. Just as we have a Man in the Moon, so they had a Rabbit in the Moon. On the bowl a star is shown at the rabbit's foot, and this seems to be the position in which the star would have been seen, relative to the crescent moon, on 5 July 1054, when the supernova was at its brightest. Twenty-three rays emanate from the star on the bowl, and the American scientists Ralph Robbins and Russell Westmoreland suggest that this represents the 23-day period when the supernova was visible in the daytime.

Archæologists have also found other burial bowls with rabbit images which evidently represent lunar eclipses. The bowls were made by the Indians of the Mimbres tribe, and indicate that considerable attention was paid to the Moon. Unfortunately the Mimbres left no written records, and they are now extinct, so that we cannot ask them what they think about it all, but there is at least a chance that they have left us a record of a colossal stellar explosion which lit up our skies nearly a thousand years ago.

60 A Swarm of Icy Planets

We know of eight planets in the Solar System, plus the Earth. Five (Mercury, Venus, Mars, Jupiter and Saturn) have been known from prehistoric times. Of the outer members of the family, Uranus is just on the fringe of naked-eye visibility and Neptune well below, while Pluto is very dim, and was not discovered until 1930. It is smaller than the Moon, and may not be worthy of true planetary status. It has a companion, Charon, about half its size.

Searches for a new planet beyond Neptune and Pluto have been going on for many years, but so far with negative results. But now a new theory, due to Dr Alan Stern of the Center for Astrophysics and Space Astronomy in Boulder, Colorado, suggests that there may be hundreds or even thousands of small, icy planets, between 600 and 1200 miles in diameter, moving round the Sun at distances of thousands of millions of miles.

Dr Stern bases his theory upon the strange characteristics of the three outermost known planets. Uranus is 'tipped on its side', so that at times one of its poles is turned toward the Sun, making the Uranian calendar very odd indeed. Neptune has a large satellite, Triton, which moves round the planet in a wrong-way or retrograde direction, in the manner of a car driving the wrong way round a roundabout, while Pluto is also tipped over, and has a strange orbit which can bring it closer-in to the Sun than Neptune. (This is the case at the present time; not until 1999 will Pluto regain its title of 'the outermost planet'.) Dr Stern believes that collisions with other bodies account for these strange phenomena. If Uranus were hit by a planet-sized body early in its career, it could have been tipped over. Another collision

Neptune and Triton, shown together in this last Voyager 2 picture in 1989. Reproduced by kind permission of NASA.

could have thrown Triton into a retrograde orbit round Neptune, and, incidentally, given a highly eccentric orbit to another of Neptune's satellites, Nereid. Finally, it has been suggested that Pluto and Charon were originally combined in one body which was shattered by yet another collision.

Three collisions, involving the three outer planets? This is highly unlikely unless there are literally thousands of potential 'colliders'. To quote Dr Stern: 'We have three smoking guns here that indicate collisions at all of these planets. Unless there is a significant number of other planetary bodies out there, the odds of even a single one of these collisions is extremely low'—less than 1 in 100,000. So according to his theory, there are many bodies which are capable of causing catastrophic collisions.

Dr Stern hopes that some of them will be found during the 1990s with the launch of the Space Infra-Red Telescope Facility. 'Instruments on this space-craft could allow for a search reaching out thirty times as far as Pluto. The further such a search is extended, of course, the better the chances for detection.'

I admit that I am dubious, if only because I am not at all sure that collision is the cause of the tilt of Uranus; it would indeed need a massive blow to tip over a globe 30,000 miles in diameter. But Dr Stern may well be right, and we must simply wait to see what future researches can tell us.

61 Astronomers and Joseph Stalin

There can be little doubt that Joseph Stalin, who ruled the Soviet Union for so many years, will go down in history as one of the most evil men of all time. Like Adolf Hitler, he was responsible for millions of deaths, and he was utterly cold-blooded about it. Astronomers were among his victims, and one of them was a friend of mine: Nikolai Kozyrev.

I first corresponded with Kozyrev when he was studying the surface of the Moon, using the large telescope at the Crimean Astrophysical Observatory. We were both interested in what are known as TLP or Transient Lunar Phenomena—mild, localized outbreaks of gas coming from below the Moon's crust—and I suggested that he should look regularly at one particular crater, Alphonsus, which is 70 miles across, and is flanked to either side by two more large walled plains (Arzachel to the south, Ptolemæus to the north). In 1958 Kozyrev not only detected an outbreak there, but was able to obtain photographic evidence. After that we were in regular touch, and when I visited the Crimea in 1959 we met for the first but by no means the last time.

Nikolai Kozyrev, pointing to a photograph of the Moon and indicating the crater Alphonsus.

Kozyrev was born in 1908. He became an astrophysicist, and went to the Pulkovo Observatory near the city which was then known as St Petersburg and has just reverted to that name, abandoning the familiar 'Leningrad'. Like many other people, Kozyrev did not get on at all well with the observatory director, Boris Gerasimovič, but he did excellent work. He continued his researches for some time, but then, on 6 November 1936, came disaster. Stalin began a purge of scientists, including astronomers, and the dreaded secret police descended upon Pulkovo. Kozyrev was only one of those to be arrested, and he was also physically assaulted. In May 1937 he was brought to trial—for what offence is not clear—and sentenced to imprisonment. After two years he was sent to a labour camp in Noril'sk. There, a fellow inmate denounced him for his scientific views; for example, he (naturally) supported the theory of an expanding universe, which ran contrary to Soviet doctrine.

Kozyrev was sentenced to ten years in gaol. He appealed, and the sentence was altered—to death! Luckily there was no firing squad in Noril'sk, and after a second appeal the sentence reverted to one of ten years in prison.

Mercifully, the case was taken up by Gregory Shain, the present Director of the Crimean Observatory. Somehow or other Shain managed to arrange Kozyrev's transfer to Moscow, in 1945, and he was finally set free on the first day of January 1947. He had been in the hands of the secret police for over ten years.

The worst was over, and Kozyrev was able to spend the rest of his life in astronomical research. Not many supported his theory of what was termed 'causal mechanics', according to which the flow of time is the main energy source of the stars—I doubt whether anyone accepts it now—but he carried out other investigations of great value, and in 1959 the American Astronomical Society awarded him its Gold Medal for his detection of volcanic activity on the Moon. It was never easy for him to leave the Soviet Union, but he was able to do so on one occasion, and actually visited me at my home in Sussex.

Kozyrev was only one of the victims of Stalin's purge. Others, including Gerasimovič, were even less lucky; they were shot. The astronomers' wives (including Kozyrev's) were also imprisoned for various periods. It is an appalling story, and takes us back to the darkest days of Communism which, thankfully, now seem to be well and truly over.

Kozyrev was a good friend and a charming companion. I am glad to have known him.

62 Make Your Own Comet!

Halley's Comet, which last passed by us in 1986, is a famous object; it has been seen regularly since well before the time of Christ, and we confidently expect it back in the year 2061. There are, of course, many other comets, and some of them have been far more brilliant than Halley's, but these 'great comets' have been depressingly rare in our own century. So what about making a comet for ourselves?

This was actually done just after Christmas 1984. Obviously it was a scientific experiment, and it has not led to a new permanent member of the Solar System, but it was decidedly exciting.

It was created from a West German artificial satellite, orbiting around 60,000 miles above the ground. While it was over the Pacific it released two canisters of barium, the metallic substance which is used in X-rays of the human digestive tract. As soon as the barium was exposed to the Sun it radiated a coloured light which looked like a comet's tail. Before long the tail had extended over a distance of 31,000 miles, and it lasted for a full quarter of an hour.

The purpose of the experiment was to measure the effects of solar wind, which is made up of a stream of atomic particles being sent out by the Sun continuously in all directions. As the solar wind particles reach the Earth, they overload the so-called Van Allen zones, which are rings of radiation surrounding the globe; particles cascade down into the lower atmosphere, and produce the lovely glows of the auroræ or polar lights. The tail of the artificial comet was soon dissipated by the solar wind particles, but it had done everything that had been hoped of it. Several nations were involved, including Britain, the United States, West Germany and Argentina; one of the aircraft monitoring the experiment was flown by the Argentine Air Force. The overall co-ordinator was Gerhard Haerendel of West Germany, who said that the experiment 'worked perfectly, on schedule', and that the tail was about six times the width of the artificial comet's head.

According to Bob Cameron of NASA, who was on board an aircraft which had taken off from Los Angeles, 'At the outset it exploded. It looked like a very bright star, a sort of yellowish-blue flash which changed quickly to purple. It held that size and shape for about three minutes, and then we began to see a pronounced tail, which grew rapidly.' Only two of the four gas cylinders were used, but they were enough. Clouds prevented observations from being made at the Kitt Peak Observatory in Arizona, but the barium tail was seen by the instruments on two satellites, one British and one American, which had been launched in August 1984 together with the West German vehicle. The results have been good, and have provided important new data about the solar wind and magnetic phenomena in general. Initially, indeed, the whole mission had been planned by the Max Planck Institute in Munich to monitor the effects of solar wind upon the Earth's magnetic field.

At least it was something new: the first-ever DIY comet!

63 The Collapsing Telescope

Disasters in conventional astronomy are, mercifully, few and far between—but one happened on 15 November 1988, when one of the world's major radio telescopes suddenly collapsed. There was absolutely no warning, and the telescope was in use at the time, so that it was very fortunate that nobody was hurt. In a matter of seconds, all that was left of one of the world's most important astronomical instruments was a heap of scrap-iron.

The Green Bank radio telescope, which collapsed so dramatically.

The telescope, at the National Radio Astronomy Observatory at Green Bank, West Virginia, was in the form of a huge dish, 300 feet in diameter. It was a very famous instrument; way back in 1960 it had been used to 'listen out' for possible intelligent signals from stars not too unlike the Sun, and it had been playing a major part in all radio astronomy investigations. So what went wrong?

Sabotage was ruled out at an early stage. The next idea involved metal fatigue. We know that this does happen, but could there be a general failure of enough metal to bring the entire dish crashing to the ground? It did not seem likely, but after almost a year an investigating committee came up with the answer. The whole collapse was due to the failure of one piece of highly stressed steel.

This piece of material was one of four gusset plates which collected forces from the dish and its supporting structure, and transferred them to the elevation bearings of the antenna. The phenomenon is popularly known as fatigue cracking, though more accurately as progressive cracking. Steel which was repeatedly stressed when the telescope was moving eventually became brittle, and cracked. Such a crack develops slowly at first, and is therefore very hard to detect; moreover, with the Green Bank telescope, paint had been added to resist erosion, and there were also several other parts of the mounting which hid the defective gusset.

There is another point, too. Modern computer calculations show that the stresses in many parts of radio telescopes of this sort are higher than was expected when the telescopes were built. Therefore, an eventual failure of some component was inevitable—and in retrospect it is not surprising that this happened at Green Bank, which had been in operation for just over 26 years. The obvious question now is what will happen with other dishes, some of which, such as those at Parkes in Australia and Jodrell Bank itself, are older than Green Bank. Clearly there will have to be extra checks.

Meanwhile, work has already begun on a replacement telescope, which will be a fully steerable 100-metre dish. It is officially known as the Green Bank Telescope (GBT) and should be ready by the end of 1995. It is being built less than a mile east of the site of the old 300-foot, and will be an offset paraboloid, so that it will have a completely unblocked aperture. The reflecting surface will consist of over two thousand panels, each of which will be automatically adjusted as the dish moves to compensate for gravitational deformation. A laser-ranging system will control the figure of the dish, and will also be used to help the telescope pointing.

By the end of 1991 the excavation was almost complete, and initial work on the foundation was well under way. Let us hope that the new GBT will not collapse in the way that its predecessor did!

C64 The Women of Venus

The surface of Venus: an actual picture sent back direct from the soft-landing Russian probe Venera 9.

Venus is a strange planet. It was named after the Goddess of Beauty, and as seen with the naked eye it is indeed glorious, shining down far more brilliantly than any other celestial body apart from the Sun and the Moon. Yet it has proved to be a fiercely hot, lifeless world, with a thick carbon-dioxide atmosphere and clouds which contain large quantities of sulphuric acid.

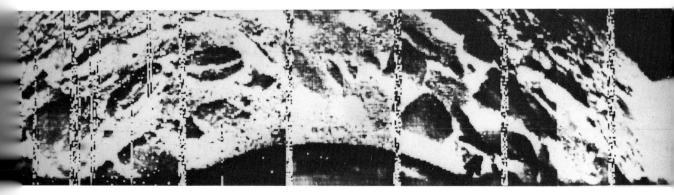

Model of Magellan. This full-scale model of the successful Venus probe is on display at the Jet Propulsion Laboratory in California.

Before the Space Age we knew little about Venus, but by now we have good maps of the surface, provided by radar equipment carried in space-craft. Venus has mountains, valleys and craters. At an early stage it was agreed that the features should be named after women, and the two main highland areas are Ishtar Regio and Aphrodite Regio, while the plains (*planitia*) have names such as Atalanta, Guinevere, Lavinia and Helen. But what about the craters revealed by the radar of the latest space-craft, Magellan, which has been orbiting Venus and sending back amazingly detailed pictures? Naming is the responsibility of the International Astronomical Union, and at the General Assembly in Buenos Aires in 1991 there were official recommendations which were approved.

There are various rules, laid down by the IAU some time ago for the nomenclature of all Solar System bodies. Duplication of names is to be avoided as far as possible, though asteroids are exempted simply because there are so many of them. Solar System nomenclature should be international; no names having political, military or religious significance can be included, and nobody can be honoured unless he (or she) has been dead for at least three years. At Buenos Aires, more than a hundred new names for Venus features were ratified.

Some of these names are interesting. For example, María Celeste is on the list; she was Galileo's daughter, and is worthy of inclusion even though the great Italian's domestic life was somewhat irregular. Nobody will question the inclusion of Joliot-Curie; here too is Woolf (Virginia Woolf, the eccentric British writer), together with Piaf (the French singer) and Marsh (Ngaio Marsh, the New Zealand writer of detective stories).

Cleopatra, one of the features on Venus recorded by Magellan's radar; it is a 60-mile depression in a hilly area. Reproduced by kind permission of NASA.

The arts are represented by Flagstad (Kirsten Flagstad, Norwegian soprano), Danilova and Nijinskaya (Russian ballet dancers), Halle (Austrian violinist) and Henie (Sonja Henie, Norwegian skater)—and it would be unthinkable to leave out Callas (María Callas, the formidable and never-to-be-forgotten Greek singer). Another newcomer is 'Monna Lisa', though her enigmatical smile is hardly reflected in her crater, and from most people's point of view her name has acquired an extra N.

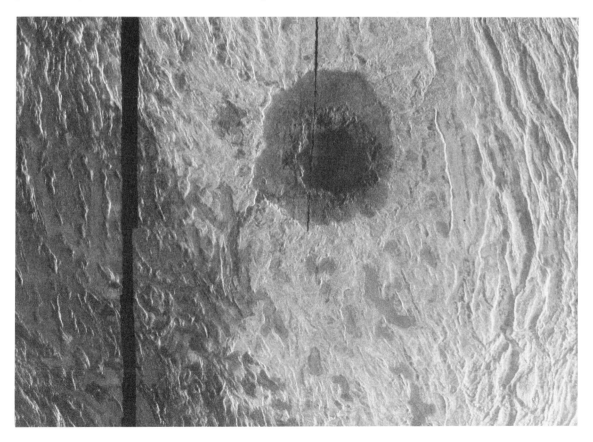

History is not neglected. Stuart commemorates Mary Queen of Scots, who came to an untimely end after a career which cannot be said to have been wholly creditable. There is probably more to be said for the Roman empress Anna Faustina, and for Julius Cæsar's mother, Aurelia. There are several deities and geographical names, and women astronomers now allocated craters include de Lalalde (France), Scarpellini (Italy) and Simonenko (Russia). We have, too, craters named after Margaret Mead, the anthropologist; Pearl Buck, the novelist; Lilian Hellman, the playwright; Rachel Carson, the environmentalist, and Clare Booth Luce. A group of three small craters has been named after Gertrude Stein, the American writer who is possibly best remembered for a remark about roses. (The other Steins are male, and do not qualify. Do you remember the somewhat unkind verse:

. . . The extraordinary family of Stein;
There's Ep, and there's Gert, and there's Ein.
Ep's statues are junk,
Gert's novels are bunk,
And nobody understands Ein.)

Most of the names are reasonably well known but there are also some others which do not strike so familiar a chord. See if you can identify the following: (a) Quetzalpetlatl, (b) Xiao Hong, (c) Al-Taymuriyya, (d) Hwang-chi, (e) Erxleben and (f) Titibu. Answers on page 128.

☾65 Stellar Neighbours

Most people know that the nearest star, apart from the Sun, is Proxima Centauri, which lies at a distance of just over four light-years (one light-year being the distance travelled by a ray of light in one year: rather less than 6 million million miles). The nearest of the really brilliant stars is Sirius, at 8.7 light-years, while the closest stars which are at all similar to the Sun, and might therefore be expected to have systems of planets, are further away— over 10 light-years.

There are many dim stars within 40 light-years of us, but most of these are dim, red stellar embers, glowing faintly and therefore well below naked-eye visibility. Recently I thought that it might be interesting to check on the lists and see just which naked-eye stars can qualify as 'neighbours'. As a limit I took apparent magnitude 4; remember that apparent magnitude is a measure of how bright the star looks, not how powerful it really is. (Canopus, in the southern hemisphere, is half a magnitude fainter than Sirius, but it is really a cosmic searchlight, perhaps 200,000 times the power of the Sun, whereas Sirius is a glow-worm by comparison.)

So here is my list. Various catalogues differ, so I have adopted that of the Cambridge Observatory, which is as authoritative as any:

Star	Name	Apparent magnitude	Spectrum	Luminosity, (Sun = 1)	Distance, (light-years)
Beta Aquilæ	Alshain	3.7	G8	3.2	36
Alpha Boötis	Arcturus	−0.04	K2	115	36
Alpha Canis Majoris	Sirius	−1.5	A1	26	8.7
Alpha Canis Minoris	Procyon	0.4	F5	11	11.4
Eta Cassiopeiæ	—	3.4	G0	5.9	19
Alpha Centauri	—	−0.3	K2+G1	1.3	4.3
Tau Ceti	—	3.5	G8	0.4	11.9
Chi Draconis	—	3.6	F7	2	25
Epsilon Eridani	—	3.7	K2	0.25	10.7
Delta Eridani	Rana	3.5	K0	2.3	29
Beta Geminorum	Pollux	1.1	K0	60	36
Zeta Herculis	Rutilicus	2.8	G0	5.2	31
Mu Herculis	—	3.4	G5	2	26
Beta Hydri	—	2.8	G1	7	21
Alpha Hydri	—	2.9	F0	8	36
Beta Leonis	Denebola	2.1	A3	18	39
Gamma Leporis	—	3.6	F6	2	26
70 Ophiuchi	—	4.0	K0	0.4	16.6
Pi³ Orionis	—	3.2	F6	3	25
Delta Pavonis	—	3.6	G5	1	18.6
Iota Persei	—	4.0	G0	2.6	39
Alpha Piscis Australis	Fomalhaut	1.2	A3	22	22
Xi Ursæ Majoris	—	3.8	G0	0.9	25
Beta Virginis	Zavijava	3.6	F8	2.7	33
Gamma Virginis	Arich	2.7	F0+F0	8	36

Of these, only a limited number are more powerful than the Sun; only one—Arcturus—has more than 100 Sun-power. When we look for really solar-type stars, we find Delta Pavonis and Xi Ursæ Majoris, both of which are about the same size, type and luminosity. Whether they have orbiting planets is another matter.

Whether we will ever reach any of these 'neighbours' is problematical. At the moment we have to admit that interstellar travel is a wild dream of the future—but then, so was reaching the Moon less than a century ago.

66 Tuttle's Comets

During August 1864, when the American Civil War was raging, a most unusual astronomer was making observations of a comet—not from an official institution, but from the deck of the Union ship USS *Catskill*, an ironclad with a single turret. The astronomer, an acting paymaster in the Union Navy, was Horace Tuttle, one of the more colourful characters in scientific history.

He was certainly brave. Apparently he was entirely responsible for the capture of an English blockade-runner ship, *Deer*, on 19 February 1865, when he went ashore and put up a signal which the *Deer* took as an all-clear sign; when the ship entered harbour, Tuttle and his men boarded it and claimed it as a prize for the *Catskill*. Tuttle was also an expert observer. He had worked as an astronomer at the Harvard College Observatory, and had discovered a grand total of four comets plus independent discoveries of another nine, including three which are periodical. Throw in a couple of asteroids, Maja and Klytia, and you have an enviable record. It is not surprising that in 1859 he was awarded the Lalande Prize from the Academy of Sciences in Paris.

Some of Tuttle's comets are of exceptional interest. One of these, known solely by his name, has a period of 13½ years. He found it during a random sweep on 5 January 1858, when it was of the 8th magnitude; subsequently it rose to 6½, so that it was not far below naked-eye visibility. Tuttle computed the orbit, and suggested that the comet might be identical with one seen in January 1790 by Pierre Méchain. The next return was fixed at 1871, and the comet duly turned up; since then it has been seen at every return except that of 1953, when its position in the sky was hopelessly unfavourable. Sometimes, as in 1980, it can be seen with binoculars, and occasionally it develops a short tail.

On 3 May 1858 Tuttle discovered a faint comet in the constellation of Leo Minor. The period was given as around 6½ years, but it was not recovered until 1907, when M. Giacobini at Nice in France discovered a comet which subsequently proved to be identical with Tuttle's. Again the comet was lost, but was recovered in 1951 by L. Kresak in Czechoslovakia, so that its official designation is now Tuttle–Giacobini–Kresak. Normally it is very dim, but at the return of 1973 it displayed a sudden flare-up which raised it to the fourth magnitude! The outburst lasted for less than a fortnight, and no explanation for it has been found.

Another of Tuttle's comets, discovered on 6 January 1866, is associated with the Leonid meteor stream; Tuttle had been anticipated by W. Tempel at Marseilles, who had first seen the comet on 19 December 1865, so that it is known officially as Tempel–Tuttle. However, there is no doubt that it had been seen earlier, in 1366 and in 1699, when it was quite bright. It was back again in 1866, the year of a great Leonid meteor storm; the

---

(a) Quetzalpetlatl: Aztec fertility goddess.
(b) Xiao Hong: Chinese novelist.
(c) Al-Taymuriyya: Egyptian authoress.
(d) Hwang-chi: 16th-century Korean poetess.
(e) Erxleben: German scholar.
(f) Titibu: Japanese Haiku poetess.
If you have scored 6 out of 6, you have done very well indeed!

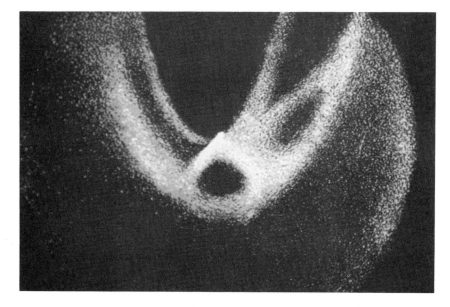

Drawing of comet Swift-Tuttle, made at its only known appearance in 1862. The detail may well be exaggerated.

period, just over 33 years, was well established, but no trace of the comet was seen either in 1899 or 1932. It was almost given up for lost, but in April 1965 it was recovered as a very dim, tailless object of the 16th magnitude. It was hoped that there might be another major Leonid display, and this duly occurred on November 17, 1966, though unfortunately it took place during daylight over Europe; observers in America were treated to a glorious meteor storm. The next return, that of 1998-9, is eagerly awaited. If we do not have another Leonid 'spectacular', we will be unlucky.

Even more interesting is Comet Swift-Tuttle, discovered on 16 July 1862 by Lewis Swift and independently on 19 July by Tuttle. It was impressive, reaching the second magnitude and developing a 30-degree tail, and was associated with the August Perseids—the main annual shower. The period was given as 120 years, and so the comet ought to have returned in 1982. Nothing was seen of it, and it may well have come and gone unseen, though there is also a suggestion that it could be identical with Kogler's Comet of 1737, in which case the period would be appreciably longer. Whether we will ever see it again is uncertain.

Now let us return to the career of Horace Tuttle himself. In 1869 he was awarded an honorary MA degree from Harvard, and was universally respected, but trouble lay ahead; it was found that there was something wrong with his Naval account books. Not to put too fine a point on it, they were 8800 dollars short. Four years later Tuttle was caught cashing cheques which were rightly the Navy's. Not unnaturally, he was charged with embezzling and 'scandalous conduct'. There was not a great deal to be said, and he was dismissed from the Navy.

Was this the end? Not at all. Only three weeks later we find him installed as Astronomer to the Rocky Mountain Region of the US Geographical and Geological Survey, and by 1884 he was back in Washington, this time working at the US Naval Observatory. He must have had a plausible tongue, and subsequently he even acted as a contractor for the United States mail service.

He does not seem to have found any more comets, but he remained active, and was an excellent author of popular articles on astronomy. Apparently he remained on good terms with the law from then until his death in August 1923. I think I can do no better than quote the words of the celebrated astronomer and historian Donald Yeomans in his classic book *Comets*: 'Horace P. Tuttle—Comet Hunter, Civil War Hero, and Embezzler!'

☾67 The Giant Sun

The Sun is an ordinary star; it is in orbit round the centre of the Galaxy, taking 225,000,000 years to complete one circuit (a period often called the cosmic year). It is around 5,000 million years old, and at the moment it is steady and well-behaved, though it does show a mild eleven-year cycle of activity.

We have long known that it is creating its energy by nuclear reactions going on near its core. The solar 'fuel' is hydrogen, and the supply is not inexhaustible, so that eventually the Sun must change its structure—with tragic effects on the Earth.

Calculations indicate that in between 1½ and 2 thousand million years from now the Sun will be 15 per cent more luminous than it is at present, so that the climate of Iceland will become rather like that of modern California. The real crisis will come in about 5000 million years' time, when there is no more hydrogen fuel available. The core will shrink, while the outer layers will expand; the overall luminosity will increase, and the Earth's temperature will rise by around a hundred degrees, so that the oceans will evaporate. Over the next cosmic year the Sun will swell even more, and the Earth's surface will become a sea of molten lava. To any surviving inhabitants, the daytime Sun would almost fill the sky.

This sounds depressing enough, but there is worse to come. Calculations made in 1988 by Dr J. Goldstein, in Pennsylvania, indicate that the red giant Sun will be so large that it will have a diameter greater than that of the Earth's orbit, so that we will be inside it. True, the outermost layers of a red giant are almost incredibly rarefied, and are cool; all the same, the temperature of the Earth's upper air is bound to reach at least 1800 degrees Centigrade, and presumably the atmosphere will be stripped away.

The orbit will change, too. It will decay—that is to say, the Earth will spiral down toward the Sun's centre. In only 300 years it will have shrunk to one-hundredth of its present size, but by then, needless to say, the world will have been destroyed. When the Sun finally collapses into a very small, dense, feeble white dwarf, preparatory to ending its life as a dead globe, there will be no Earth-dweller to see it—unless something can be done.

We cannot modify or halt the Sun's evolution, but if mankind still survives 5,000 million years hence we will know a great deal more than we do now,

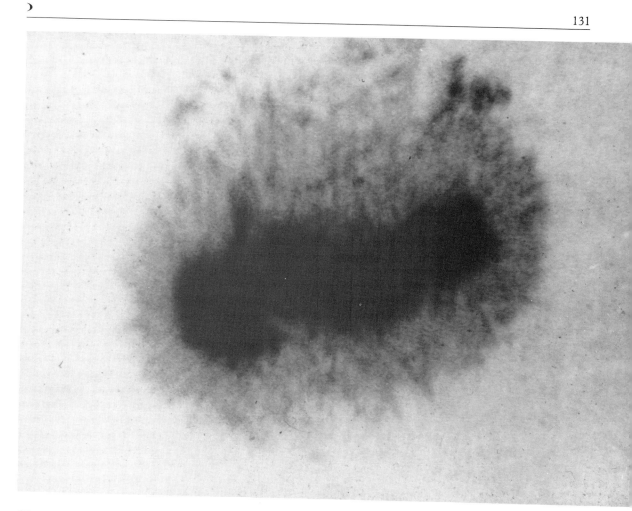

Giant sunspot.

and it may not be impossible to either move the Earth to a place of safety, or else create an energy source for ourselves which does not involve the Sun at all. After all, it has taken *homo sapiens* only a few thousands of years to progress from stone axes to nuclear reactors. Probably the main danger is that unless we learn how to govern ourselves more sensibly, we may well destroy the entire human race before the Sun does.

68 Space Bureaucrats

Space research can provide moments of hilarity, dangerous and daunting though it may be. There have been amusing episodes, and one of these, dating from 1989, concerns the Russian space-station Mir.

Mir is to all intents and purposes a permanent manned base in orbit, and it is quite unlike America's Skylab, which was never intended to remain aloft for a long period even though it had a useful lifetime of several years. Mir

Colonel Romanenko, one of the cosmonauts who has spent many months on the Mir space-station. This was a picture of him which I took in 1990 when he was on a visit to London.

has been a major success. Almost all the time since it was launched it has carried Soviet cosmonauts, as well as visitors from other nations (one Briton), and an immense amount of scientific work has been carried out. Cosmonauts who have been there have a genuine affection for it. Not long ago I had the pleasure of talking to Colonel Yuri Romanenko, who was on the station for almost a year and has suffered no ill-effects (though it is true that his original space-flight engineer had to leave half-way through the mission because his heartbeat showed signs or irregularity). He is quite optimistic about flights to Mars, and does not believe that the human body will be too frail to stand up to such a journey.

Another of the Russian cosmonauts is Sergei Krikalev. Cosmonaut selection in the USSR (or what used to be the USSR!) is every whit as rigorous as it is in the United States; training is very hard indeed, and only those capable of dealing with any emergency have any hope of passing through the whole course. In general they range in age between the mid-twenties and the mid-forties, and they come from all walks of life.

I have not met Sergei Krikalev, but from all accounts he is a thoroughly efficient member of the team, and has played his part in various important projects. But he is also of military age, and in Russia, as in most countries, it is compulsory to do a period of military service. So in March 1989 Krikalev was sent written orders to report to the district Army draft centre. In fact there were several copies of the order, all duly delivered.

There was just one problem. At that moment Krikalev was on board the Mir space-station, and had been there for four months!

In fact, his name and those of his two companions—Alexander Volkov and Valeri Polyakov—had been mentioned on Soviet television almost every night, but the Moscow bureaucrats clearly had not noticed. It seems that civil servants are the same the world over. . . .

Krikalev did not dash back to obey his draft call, and whether he reported for military service on his return to terra firma I do not know. But it is, surely, yet another indication that there are many officials who need their heads examined.

☾69 The Glatton Meteorite

Glatton, a little village in the English countryside not far from Peterborough in Cambridgeshire, is not the sort of place which would be expected to hit the astronomical headlines. Yet on 5 May 1991 it did so, with a vengeance.

Mr Arthur Pettifor, who is eighty years old and a retired civil servant, was quietly tending his garden when he was rudely disturbed. 'I heard a very loud whistling, whining noise over my head', he said. 'It was very alarming. Then I heard a loud thud, and I saw the top of a conifer tree shaking where it had been hit. I thought that something had dropped from an aeroplane; then I found a lump of black rock under the hedge.'

What Mr Pettifor had found was nothing more nor less than a meteorite. It was only about three inches in maximum diameter, weighing 1½ pounds, but its meteoritic nature was unquestionable, and Mr Pettifor very sensibly reported it. Before long the experts arrived on the scene. The meteorite was taken to the Natural History Museum for analysis, and was found to be a chondrite—that is to say a stony object, with small pieces of mineral (chondrules) embedded in it. Olivine and silicates are the main constituents, but there are traces of iron and nickel as well.

No sightings apart from Mr Pettifor's have been reported, but it looks as though the object must have streaked across the English sky at about 12.30 p.m. on Sunday May 5. For a few seconds it could well have outshone the full moon, but there was a good deal of cloud around, and I doubt whether there were any further sightings.

There seems no doubt that meteorites come from the asteroid belt, and falls are not unusual, so that on the whole it seems rather strange that there have been no human casualties. The Glatton Meteorite missed Mr Pettifor by about twenty yards. But for every genuine meteorite reported there are many false alarms, and I was involved in one of these not so long ago.

The story began on 14 February 1989, when several British newspapers carried a story about a meteorite which had come down at Lewes, in Sussex, not far from the old site of the Royal Greenwich Observatory at Herstmonceux. Apparently a small iron meteorite, about two inches across, had crashed through the roof of Lewes Railway Station. Dr Andrew Graham, of the Natural History Museum—a world authority on meteorites—was quoted as saying that it was 'very exciting, and the first meteorite seen to fall in Britain for twenty years' (the last had been the Bovedy Meteorite, in Northern Ireland, in April 1969).

The Glatton Meteorite. Reproduced by kind permission of Jonathan Shanklin.

I was dubious from the start, because Lewes is a chalky area, and nodules in chalk can look very like meteorites. However, it was clearly worth finding out, so I got into my car and drove the seventy miles to Lewes, where I interviewed the stationmaster, Mr David Cates, who had seen the fall and was quoted in the papers as saying that he planned to use the meteorite as a paperweight.

A few minutes' conversation with Mr Cates cleared the whole matter up. Let it be said at once that he acted with very sound common-sense throughout. To anyone who is not an astronomer or a geologist the sudden arrival, via a glass roof, of an object with a brown exterior and roughly circular shape does indeed suggest a meteorite, so Mr Cates took advice, with the inevitable result that he was bombarded by inquiries; the local Press arrived, and before long the story reached the national papers.

Mr Cates telephoned a geologist friend, who told him to take the object outdoors and hit it with a hammer. He did, and all was revealed. It later transpired that the object had been dug up from a flowerbed some days earlier, and had been thrown around by boys at play, which is how it arrived through the station roof.

What about Dr Graham? What he had actually said was that *if* the report of a meteorite turned out to be true, it would be exciting. Of course he was quite right, but if Mr Cates had acted with less intelligence a great deal of time would have been wasted in trying to track down a non-existent meteorite. It shows how careful one must be.

☾70   The Faintest Galaxy

Our Sun is a member of the Galaxy, which is made up of about 100,000 million stars, and is a flattened system, so that when we look along its main plane we see many stars in almost the same direction—producing the familiar Milky Way appearance. The Galaxy is one of several making up what we call the Local Group. In this Group the two brightest members, in our skies, are the two Clouds of Magellan, both of which lie within 200,000 light-years of us; unfortunately for northern-hemisphere observers they lie in the far south of the sky, but they are almost ever-present in the night skies of countries such as Australia, New Zealand and South Africa. Both are much smaller than our Galaxy, but their relative nearness makes them prominent, and to the naked eye they look rather like broken-off portions of the Milky Way.

However, the Local Group contains at least one system, the Andromeda Spiral, which is larger than ours, and contains at least 150,000 million stars of all kinds. At its distance of rather over 2,000,000 light-years it is just visible without optical aid if you know where to look for it; binoculars show it easily, and photographs taken with adequate telescopes show that it is a spiral, though because it lies at a narrow angle to us the full beauty is lost. Not far from it is a fainter system, the Triangulum Spiral. The Local Group also includes

Messier 31, the Andromeda spiral, a huge star system 2,200,000 light years away. It is larger than our own Galaxy and is just visible with the naked eye. It is spiral but lies at an awkward angle to us.

more than two dozen much smaller galaxies, known as dwarfs, most of which are irregular in form.

One dwarf, in the constellation of Carina (the Keel of the Ship) is regarded as a satellite galaxy of the Milky Way system; it was found in 1977. Much more recently, in 1990, a new satellite galaxy has been identified by Michael Irwin, Richard McMahon, Peter Bunclark and Mick Bridgeland, working at the Cambridge Observatory and using plates taken by the Schmidt telescope at Siding Spring, in Australia. The new dwarf is in the constellation of Sextans (the Sextant) and is estimated to be about 300,000 light-years away—further than the Clouds of Magellan, but still on our door-step by cosmical standards.

Like the Carina dwarf, it is spherical. The total luminosity of all its stars combined is about 100,000 times than of the Sun. This is less than a really powerful single star in our Galaxy, such as Canopus, which according to the authoritative Cambridge catalogue could match 200,000 Suns, or two Sextans galaxies! This, of course, is why the dwarf has not been located before.

The movements of these satellites of our Galaxy can give vital information about the masses of the systems, and it turns out that there is much more mass than can be explained by visible stars. In fact, there is a tremendous amount of dark material which we cannot see. The Sextans dwarf has a mass equal to about ten million Suns, and contains hundreds of millions of individual low-mass stars, but there is more to it than that.

Like other dwarf satellites, it may have lost much of its interstellar gas when it was 'captured' by the Milky Way; but from whence did it come? Is it a fragment left over from the formation of a large galaxy, or is it the débris produced when two galaxies collided? We do not yet know, but at least the Sextans dwarf seems to be the least luminous system yet identified, and as such it is definitely an important 'find'.

ℂ71 Jim Irwin

A few months ago an old friend of mine died. His name was Jim Irwin—more officially Colonel James Benson Irwin, late of the United States Air Force. In July 1971 he became the eighth man to step on to the surface of the Moon, so that he belonged to a very exclusive club indeed.

He was born in Pittsburgh on 17 March 1930, and took degrees in naval sciences, aeronautical engineering and instrumention engineering, qualifying as a pilot. He joined NASA in 1966, and became involved in the Apollo programme of lunar missions, though, as he told me on one occasion, he was never sure in those early days at NASA whether he would ever make an actual flight.

The idea of sending men to the Moon goes back a long way, and in the second century AD a Greek writer named Lucian wrote a story about a Moon voyage (though he certainly did not expect it to be taken seriously). In the 19th century came the famous novel by Jules Verne, in which the travellers were fired moonward from the barrel of a huge gun. But it was only in our own time that travel to the Moon became practicable. The Apollo programme was initiated in the early 1960s; in 1969 Neil Armstrong became the first man to tread lunar soil, from the *Eagle* module of Apollo 11, and then came Apollos 12, 13 (a failure), 14, and then 15, with David Scott and Jim Irwin.

I was one of those who had been mapping the Moon in the pre-Apollo days, and during the actual voyages I was carrying out the commentaries for BBC television. I was able to hear the voices of Dave and Jim as they set up their equipment, and then drove around among the lunar craterlets and the foothills of the Lunar Apennines in the Moon-car that they had brought with them. They knew exactly what they had to do, and they carried out their mission with absolute precision. When they had completed their full programme they blasted off, re-joined their companion (Alfred Worden) in the orbiting section of Apollo 15, and came home.

Put that way, it may sound straightforward enough—but it was anything but easy. Space is a dangerous environment, particularly when you are a quarter of a million miles away from home. Dave Scott and Jim Irwin had to depend not only upon the technology of those who had sent them to the Moon, but also upon themselves. They were equal to the occasion; otherwise, they would not have been astronauts.

Two more Apollos followed No. 15; No. 16 came down in the southern highlands, and finally No. 17, commanded by Eugene Cernan, took a

Jim Irwin on the Moon, with Apollo 15, saluting the US flag. Reproduced by kind permission of NASA.

professional geologist, Dr Harrison Schmitt, to the lunar surface. But Apollo 15, with its Moon-car, had shown the way.

Jim Irwin remained with NASA until 1972, though he made no more trips into space. After the end of his career as an astronaut he did a great deal of public speaking, and became something of an evangelist, which led to some Press speculation that his trip to the Moon had caused him to alter his whole outlook on life. This is not true. Religion always played a great part in his way of thinking, and he never changed. The last time I saw him was at Expo 88 at Brisbane, in Australia, where I had the honour of being co-speaker with him at the main meeting. He was above all a modest man; it was not always easy to credit that he was one of the select few who have walked on the surface of another world.

I have a slightly poignant memory of that meeting. When the main ceremony was over, both Jim and I were presented with tungsten wrist-watches similar to those worn by some of the astronauts. The strap of my watch was too small for my wrist; Jim's was too large for his, so that a few links were suitably transferred. The watch-strap that I am wearing as I type these words therefore contains some of the links from Jim Irwin's.

He died, suddenly, of a heart attack, and is in fact the first of the Moon-men to leave us. Believe me—and I am in an excellent position to judge—he was a great man as well as a great explorer. He is much missed, and he will not be forgotten.

☾72   The White Spot on Saturn

Saturn, the second of the giant planets of the Solar System, is noted mainly because of its superb system of bright rings. Observers often neglect the yellowish, flattened disk, which shows darkish belts and bright zones, similar to but much less impressive than those of Jupiter, and occasional spots and festoons. Sometimes, however, there are violent outbreaks, and one of these was seen in 1990.

The story began on 24 September, when two American amateurs, Stuart Wilber in Las Cruces and Alberto Montalvo in Los Angeles, turned their telescopes toward Saturn and noted a brilliant white spot just north of the planet's equator. Because Wilber reported it first, he is officially credited as the discoverer. At that time, unfortunately, I was abroad at a conference and therefore away from my telescope, but as soon as I came home almost the first thing I did was to look, using my 15-inch reflector. Sure enough, there was the white spot. It was just as bright as the spot I remembered seeing in 1933, which had been discovered by W. T. Hay (better remembered today as Will Hay, of stage and screen fame). It was much more conspicious than the white spot of 1962.

Montalvo's description of the new outbreak was 'a bright white spot on the disk. It was like a flashlight, five or six times brighter than the rest of the planet'. It seems to have been due to an eruption from below Saturn's outer clouds which had swept crystals of ammonia into view; these crystals had not had time to be turned yellowish by the action of sunlight, so that the spot was clearly seen.

It remained an oval patch for about three weeks, but by October the storm centre showed brilliant spots, and developed a tail extending eastward. This was predictable, because the eastward winds on Saturn are the strongest in the Solar System. They can blow at up to 1000 miles per hour, and a 'cloud' can go right round the planet in 10 hours 15 minutes, whereas Saturn's core has the longer rotation period of 10 hours 39 minutes.

The spot was literally torn apart, and slowly its identity was lost. It turned into a bright band round the planet's equator, though part of its material, at the source latitude of about 4 degrees north, remained distinct for some time. By November, when a superb picture was taken with the Hubble Space Telescope, the collision between the strong equatorial winds and the gentler mid-latitude winds had distorted the northern edge of the band into waves and eddies, so that the equatorial zone became bright and featureless. The Great White Spot, as such, was no more.

It was important because it provided a natural 'marker' from which we could measure the rotational period of Saturn's outer clouds. When another comparable spot will appear we do not know, but they seem to happen every 29 years or so, and of course Saturn takes 29½ years to go round the Sun. Observers will be on the alert around the year 2019—and if Saturn puts on another display, it may well be that, as before, the discovery will be made by an amateur astronomer.

Great White Spot on Saturn

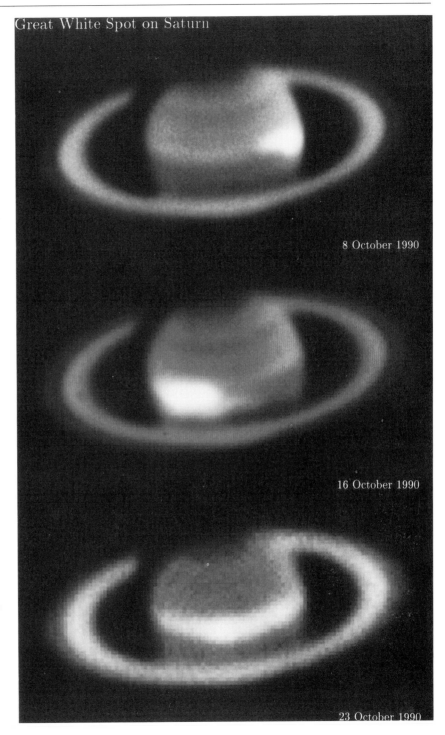

8 October 1990

16 October 1990

23 October 1990

The White Spot on Saturn. These three photographs were taken from the ESO observatory at La Silla in Chile, and show how the spot developed over two weeks. Between 8 and 16 October it grew, but after 16 October rapidly expanded to reach all the way round the planet's equator. The telescopes used were the NTT (first and second pictures) and the 2.2-metre (third picture), all with CCDs. Reproduced by kind permission of the ESO.

C73 John Herschel and the 'Solitary One'

Alphard or Alpha Hydræ, the only bright star in the huge, sprawling constellation of the Watersnake, is very isolated in the sky. There are no prominent stars anywhere near it, and it has long been nicknamed 'the Solitary One'. It is easy to find, if only because it lies in line with the famous Twins, Castor and Pollux; its official magnitude is 1.98, almost exactly equal to the Pole Star, and it is decidely red. Even naked-eye observation shows the colour, and in binoculars it is striking. The spectral type is K3, the distance 85 light-years, and the luminosity 700 times that of the Sun.

Some curious observations were made of it in 1838 by no less a person than Sir John Herschel, who had spent some years at the Cape of Good Hope, making the first really detailed survey of the far-southern stars which never rise over Europe and had therefore been neglected. Herschel was sailing back to England, and while on board his attention was drawn to Alphard, which was not regarded as a variable star. Herschel's records are as follows:

21 March. Alpha Hydræ inferior to Delta Canis Majoris, brighter than Delta Argûs (now known as Delta Velorum) and Gamma Leonis.
7 May. Alpha Hydræ fainter than Beta Aurigæ, very obviously fainter than Gamma Leonis, Polaris or Beta Ursæ Minoris. 'Though low, yet it is now a decidedly insignificant star', wrote Herschel. 'Moon, and low altitude, but it leaves no doubt in my mind of the minimum being nearly attained'.

John Herschel, who first made detailed surveys of the far-southern stars. The telescope shown here is a reflector with a focal length of 20 feet; he took it with him to the Cape in the 1830s and used it for most of his work there.

8 May. 'Alpha Hydræ still high (30°) and quite free of all cloud and haze in a very fine blue sky . . . is very decidedly inferior to Gamma Leonis . . . I incline to place the minimum of last night, as Alpha is tonight brighter than Beta Aurigæ at certain intervals of its twinkling . . . On the whole after many comparisons Alpha Hydræ is rather inferior to Beta Aurigæ.'

9 May. Alpha Hydræ inferior to Gamma Leonis but . . . evidently on the rapid increase . . . The minimum is certainly fairly passed and the star is rapidly regaining its light.

10 May. Alpha Hydræ much inferior to Gamma Leonis, rather inferior to Beta Aurigæ. It is still about its minimum.

11 May. Alpha Hydræ brighter than Beta Aurigæ no doubt; Beta much higher.

12 May. Castor and Alpha Hydræ nearly equal.

Herschel docked in London on 15 May, and apparently never made any further observations of the star.

Modern magnitude measurements make Alphard 1.98, Beta Aurigæ 1.90, Delta Velorum 1.96, Beta Ursæ Minoris 2.08, Polaris 1.99 (very slightly variable), Gamma Leonis 1.99 and Castor 1.58. If Herschel's observations had been accurate, it would follow that Alphard is variable by about half a magnitude—but so far as I know the observations have never been confirmed, and Alphard is an awkward star to estimate simply because of its isolation. To be candid, I doubt whether there is any marked fluctuation, but John Herschel's shipboard study is not without interest as an historical footnote.

74 The Curious Case of BL Lacertæ

Did you know that there is a lizard in the sky? I can assure you that there is: Lacerta, one of the original 48 constellations listed by Ptolemy, the last great astronomer of Classical times. It lies in the far north, adjoining Cepheus. It is very obscure, and has no bright stars, but it does include one fascinating object, known as BL Lacertæ.

It is too faint to be seen with the naked eye, or even binoculars, but it was noted as long ago as 1941, and was found to change in brightness, so that it was classed as a variable star (of which there are many in the sky). As a matter of routine its spectrum was examined. When a star's light is split up by a prism, or some equivalent device, the normal result is a bright rainbow band, from red at one end to violet at the other, crossed by dark lines which show the presence of some particular element or group of elements. But when astronomers tried this method with BL Lacertæ, they had a rude shock. There were no lines at all, and so, quite obviously, BL Lacertæ was not a star.

The next step was to see if it were associated with any cloud of dust or gas. This was found to be the case, and in 1973 astronomers at Palomar, in California, detected a surrounding 'fuzz' which did show some spectral lines. The way in which these lines were positioned showed that the object was racing away from us at a rate of over 12,000 miles per second. This recessional velocity gave a key to its distance, and there could be no doubt that BL Lacertæ lay far beyond the confines of our Milky Way galaxy—so was it a galaxy in its own right? This seemed reasonable, but the spectrum was quite unlike that of a conventional galaxy, which is made up of thousands of millions of stars.

It was then proposed that there might be a link between BL Lacertæ and others of the same kind—of which several were soon found—and quasars. A quasar is very remote and super-luminous, and is now believed to be the core of a very active galaxy which is powered by a massive black hole deep inside it. The latest idea is that there is no real difference between the two classes of object, but we see a quasar at a reasonable angle, or even broadside-on, while with BL Lacertæ we are looking 'straight down' a jet. This means that we see the jet as a bright point which overpowers the dim galaxy surrounding it. In fact, everything depends upon our angle of view, and by now over a hundred BL Lacertæ objects have been found.

Most of these lie in elliptical galaxies, but in 1991 Ian McHardy and his colleagues at Southampton University have found one of them in a 'flat' or disk galaxy, which is surprising; most BL Lacertæ objects are strong emitters at radio, infra-red and X-ray wavelengths, and these emitters are rare in disk galaxies. A great deal remains to be found out, and though we believe that we are on the right track there are still many mysteries associated with the curious object in the celestial Lizard.

75 Astronomers versus Squirrels

Astronomers and environmentalists do not always see eye to eye. Optical astronomers are being badly affected by light pollution, while radio astronomers are having problems with commercial and general 'noise'. The latest difficulties come, believe it or not, from squirrels!

One of the main astronomical centres of the world is Tucson, in Arizona. It ranks with centres such as the top of Mauna Kea, in Hawaii, and the Atacama Desert of Chile, where we find the main observing station of the European Southern Observatory as well as two major American observatories. Nothing much lives on the top of Mauna Kea, and I must say that the Atacama Desert is one of the most barren places that I have ever visited, with no inhabitants apart from a few donkeys and desert foxes (I have never been able to find out what they eat). But Tucson needs a good observing site, well away from the city, and the authorities of the University of Arizona settled on Mount Graham, which is reasonably high and is part of the Colorado National Forest. They planned to spend two hundred million dollars on a major observatory on the summit, involving several large telescopes and occupying a large area.

This seemed a good idea; there was no light pollution or smoke, and the skies are generally clear. But the University had reckoned without the Mount Graham red squirrels, which are very rare. What apparently happened was that at the end of the Ice Age, about 10,000 years ago, Mount Graham was connected with the Rocky Mountain Forest, but when the world warmed up, and the lowland became a desert, Mount Graham was cut off—and cut off with it were the red squirrels. (If you want to know their official name, it is *Tamasciurus Hudsonicus Grahamensis*.) There are only about a hundred and eighty of them now, and there are none anywhere else in the world, so that they are definitely an endangered species. Their habitat is the spruce-fir forest, below which, on the lower slopes of the mountain, are conifers.

Up to now the red squirrels have been left in peace—and so when the University of Arizona announced its plans for a new observatory, the environmentalists swooped. The astronomers promised faithfully that they would not hurt the squirrels, but the environmentalists were sceptical. On 26 March 1990 there was a Court hearing, and Judge Alfredo Marquez ordered that all work on the project should be stopped for a hundred and twenty days, so that the objectors could take another look and see whether they were still worried about the squirrels' survival. Astronomers felt that they were being literally driven nuts.

Of course, there are two sides to every question (one's own, and the wrong one), and if the observatory had really posed a threat to the rare squirrels some serious re-thinking would have had to be done. Everything hinged upon just how much room the squirrels wanted. Therefore, the United States Forest Service decided to undertake a careful and comprehensive study of the area, and when they issued their report it seemed that the astronomers had been right.

It had been claimed that the squirrels lived only in a definite type of spruce forest at or above a height of 10,000 feet, but this is not so. After surveying a much larger area of the mountain than had previously been done, the Forest Service biologists found that the squirrels roam over a zone of mixed conifers lying between 9,000 and 10,000 feet. Moreover, Mount Graham contains 11,000 acres of habitat for the squirrels, and more than 60 per cent of them would live outside the 1700-acre reserve containing the proposed observatory. So it now seems that the matter will be resolved amicably, though whether red squirrels will be welcome in the Observatory domes seems to be rather dubious.

76 A Hot Spot on Betelgeux

Betelgeux, in the constellation of Orion (the Hunter), is one of the most famous stars in the sky. It lies in the upper left of the Orion pattern (assuming that you are observing from the northern hemisphere of the Earth), and it is clearly orange-red, contrasting sharply with the pure white of Orion's other leading

Old figure of Orion. Betelgeux marks the Hunter's upper shoulder; Rigel, his foot.

star, Rigel. Betelgeux is a huge red supergiant. Its distance is given in the Cambridge catalogue as 310 light-years. Other estimates increase this to around 500 light-years. In any case it is very distant, and must be at least 15,000 times as powerful as the Sun.

Our ideas about stellar evolution have changed dramatically over the past hundred years or so. Originally it was believed that a red supergiant must be young, still condensing out of the gas and dust from which it was formed. Now we know that this is not true. Betelgeux and others of its kind have used up most of their nuclear 'fuel', and are drawing on their reserves, so that they are well advanced in their life-stories.

If you could go close to Betelgeux you would see that the edge would be fuzzy rather than sharply defined. This is because the star has blown out its exterior layers, leaving a smaller but unimaginably hot core with a vast 'atmosphere'. The gravity can only just hold on to the outermost layers, and eventually they will drift away, so that for a while Betelgeux will turn in to what we call a planetary nebula—a bad name, because a planetary nebula is not a genuine nebula and has absolutely nothing to do with a planet. This is by no means the final stage. Betelgeux is much more massive than the Sun, and when disaster overtakes it there is every chance that it will explode as a supernova. This is not likely to happen yet awhile, but it is a distinct possibility in a few millions or tens of millions of years.

Betelgeux has an apparent diameter larger than that of any other star, apart from the Sun. Even so, it looks like a speck of light, and it is very remarkable that astronomers using the great William Herschel telescope at La Palma, in the Canary Islands, have been able to detect a hot spot on it. Remember, Betelgeux is big enough to engulf the whole orbit of the Earth round the Sun, but its apparent diameter is no more than 50 milliarcseconds.

Some time ago it had been suggested that the outer layer of a red supergiant should consist of a few vast convection cells, where energy bubbles up from below. At La Palma, John Baldwin and Peter Warner decided to test this. Atmospheric blurring in the Earth's atmosphere is the main hazard, so they covered up the main mirror of the William Herschel telescope with a mask, leaving a few holes. This meant that the light from Betelgeux was collected and analysed from only a few patches of the mirror. These isolated beams then re-combined to form interference patterns, and a complicated computer program made it possible to obtain an image of the star, much as radio astronomers do in the method of interferometry.

Early results indicated that Betelgeux might have a small companion star close by it, but Baldwin and Warner found that this was not so. Instead, Betelgeux has a bright patch toward one edge of the tiny disk, and presumably this is one of the convection cells which theory had predicted.

This is a really tremendous achievement, even though it is very preliminary, bearing in mind that its apparent diameter is about the same as that of a three-foot ruler laid out on the surface of the Moon. Its name, incidentally, means 'the Armpit of the Central One', or Orion's shoulder. There are several ways of spelling it, and you can pronounce it almost as you like; 'Bettle-gurz' is the favourite way, but some people prefer 'Beetlejuice', which is certainly easier to remember!

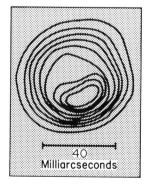

40
Milliarcseconds

C77 The Astronomical Admiral

Sailors navigate by the stars, but how many sailors have also become eminent astronomers? One who certainly did so was William Henry Smyth, who was born in London on 21 January 1788. He joined the Navy, and fought against the French during the Napoleonic Wars; then, between 1817 and 1824, he made a careful survey of the Mediterranean for the British Admiralty. He had a distinguished career, and before he retired from the Service he had reached the rank of Admiral.

This is praiseworthy, of course, but he would hardly be remembered today for his naval exploits alone. Smyth's fame came not as a Services officer, but as an amateur astronomer. He went to live in Bedford, and there set up an observatory, equipping it with a fine $5^9/_{10}$-inch refractor, and concentrating largely upon measurements of double stars. Eventually he published his great book *Cycle of Celestial Objects*, which is a survey of the entire sky. Volume I (Prolegomena) is an account of contemporary astronomy, with a description of his private observatory; Volume II is the Bedford Catalogue itself. The book became immensely popular, and was regarded as a standard work. Following its publication Smyth was awarded the Gold Medal of the Royal Astronomical Society, and served as President for two years.

Obviously, some of the sections of the book are now out of date. Though he accepts the fact that the Sun is a star, he writes that 'Many are glad to join in the opinion, that the Sun is not a globe of fire, from unwillingness to believe that a mass containing so vast a portion of the matter of the whole system cannot be habitable and inhabited. That belief is strengthened by the production of heat and fire in many ways besides through the power of the Sun; and even in those cases in which heat is supposed to be received from the Sun, the matter may be only extricated, as it were, by his influence, from substances in which it previously existed'. The red colour of Mars is 'imputed to the density of his atmosphere, which is considered as being of great extent'. The Cycle is fascinating reading, not least for the historical notes; for example Smyth quotes the 16th-century Leonard Digges, who gave the distance from London to the star Sirius as exactly 358, 463½ miles!

In a later book, *Sidereal Chromatics*, Smyth elaborated on the colours of double stars. For example the components of Alpha Herculis were 'cadmium yellow, greenish'; 95 Herculis 'light apple green, cherry red'; Gamma Delphini 'yellow, light emerald'; Psi Aquarii 'orange tint, sky blue'; Vega 'pale sapphire, smalt blue', and so on. (Incidentally, the optical companion of Vega is below the tenth magnitude, so that it would be none too easy even to see in Smyth's telescope.) He also discussed nebulæ, in which he 'noticed pale tints of white, creamy white, yellow, green and blue'. At that time the nature of the objects we now know to be gaseous nebulæ was still uncertain, and it was sometimes thought that they were simply star-clusters too remote for the stars in them to be seen individually. In 1845 the Earl of Rosse completed his 72-inch reflector at Birr Castle in Ireland, and studied the nebulæ as well as being the first to detect the spiral forms of the objects we now call galaxies. All this led to a somewhat testy comment from Smyth:

'It will be recollected by all who are really concerned about the matter, that, when the wondrous revelations of Lord Rosse were communicated to the public, certain buzzing popinjays, who hang about and obstruct the avenues to the temple of science, vociferously proclaimed that the Nebular Theory had received the *coup de grâce*. . . . Now this crude conceit was assuredly not imbibed from his Lordship's statement'. Smyth followed this by quoting a letter from his friend William Huggins, who had just made the first spectroscopic observations showing that the nebulæ really are gaseous, and are quite unlike the spiral galaxies.

Smyth died in 1865, universally respected. Then, in 1879, came a bombshell from an English astronomer named Herbert Sadler. It was devastating; Sadler published an attack on Smyth's double star measurements, claiming that they had been faked.

Faking observations is the worst crime that can be committed by any scientist, professional or amateur. Smyth was the last sort of person who would be expected to stoop to anything of the sort, and yet Sadler's attack did seem to have justification. There was an immediate outcry. Smyth's many friends were anxious to clear his reputation, and they called in S. W. Burnham, an American who was regarded as the most expert double-star observer of the time. Were Smyth's results genuine, or not?

Burnham found the answer. In many cases Smyth had given estimates for wide double-star measurements, and these estimates were often faulty simply because Smyth had failed to explain just what he was doing—and Sadler had jumped to the conclusion that the estimates were meant to be regarded as exact measurements.

The results were satisfactory inasmuch as Smyth was shown to be completely honest, and there was absolutely no attempt to falsify his results. On the other hand, his estimates left a great deal to be desired. The final judgement must be—read the *Cycle* by all means; enjoy it, as I do; but be careful about taking some of the results too seriously. In any case, Smyth deserves to be remembered as one of the truly great 19th-century popularizers of astronomy.

☾78 Do You Want to Buy a Star?

One of the nastiest money-making schemes in my experience came to my attention in 1987, when I received a letter from a man whose young daughter had been killed in a road accident. He had, he wrote, been told that if he paid a sum of £25 to an organization called the International Star Registry he could have a star named in her memory. What did I advise?

I looked into it—and was appalled.

First, let us see what is the true situation. All the stars have official catalogue numbers, at least down to a certain limit of faintness. The brightest stars, such as Sirius and Rigel, have names of their own, and the main stars in any constellation also have Greek letters; thus Sirius, the brightest star in Canis Major (the Great Dog) is Alpha Canis Majoris. All the proper names are very old, mainly Arabic, and are not much used except for stars of the first magnitude; the Greek letters were allotted by a German astronomer named Bayer in 1603. The catalogue numbers are the responsibility of the International Astronomical Union, the controlling body of world astronomy.

But wait! The International Star Registry says that if you pay your money you can choose a star and have it named anything you like, from Fred to Julius Cæsar. It is said that 'the name is permanently filed in our Swiss vault and recorded in a book which will be registered in the copyright office of the Library of Congress in the United States of America' . . . and so on. The idea seems to have been worked out in 1979 by a Canadian named Ivor Downie. Subsequently other American companies appeared, all offering to name stars, and by now about half a million have been sold. The scheme spread to Britain, and the new organizations took out advertisements in various papers, including the *Sunday Times*.

It goes without saying that the whole scheme is not only unofficial, but completely bogus. The 'names' will never be seen by anyone, and certainly not used. Some previous ventures have been amusing; for example, there was one Japanese firm which went so far as to sell fishing rights on the Moon, which presumably deceived nobody at all—but there is nothing amusing about the organizations selling star names, because they prey on people who have been bereaved.

Efforts have been made to alert prospective victims to the actual position. These efforts have met with a certain amount of success, but unfortunately the practice continues, and I still have letters about it.

So if you come across anyone who is thinking of buying a star name, do please point out that spending money in such a way is rather less useful than buying a fishing licence for the Moon.

79 The Strange Case of Zibal

Most people have heard of Sirius, Canopus, Betelgeux and Rigel, but relatively few people have heard of the star Zibal. It is not bright; its magnitude is 4.8, which means that any appreciable mist or sky-glow will drown it. Neither has it any special feature to mark it out. Its official name is Zeta Eridani, in the long, sprawling constellation of the River, not far from Orion. It lies between two rather less obscure stars, Epsilon and Eta Eridani, one of which—Epsilon—is among our nearest stellar neighbours, and seems to be a promising candidate as a planetary centre, though as yet we have no proof.

Zibal is thought to be 52 light-years away, and to have about 15 times the luminosity of the Sun; its spectral type is A, which means that it has a fairly hot surface (about 10,000 degrees Centigrade). Its main point of interest is that it was once said to be considerably brighter than it is now. Of course, we do know of many variable stars, ranging from the long-period Mira Ceti

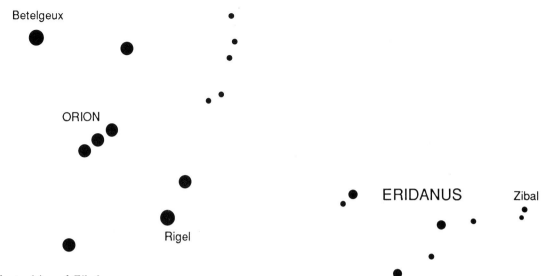

The position of Zibal.

down to the Cepheids, with their shorter periods and regular-as-clockwork behaviour; but with Zibal, what has been suspected is permanent or secular change. There is no evidence of any measurable fluctuations at the present time.

We have to be careful, because until comparatively recently we had to depend upon catalogues drawn up by observers using the naked eye only. This was the case with Ptolemy of Alexandria, the last great astronomer of antiquity, who lived around AD 150; also the accurate Arab, Al-Sûfi, around 960, and of course the eccentric Dane, Tycho Brahe, who drew up a superb star catalogue between the years 1576 and 1596 (and, *en passant*, was the astronomer who had a false nose, kept a pet dwarf, and equipped his observatory with a prison). Then came the telescopic observers such as John Flamsteed, Britain's first Astronomer Royal, who made his observations near the end of the 17th century. The point here is that Zibal was almost always recorded as being of the third magnitude, and Ptolemy made it definitely brighter than Kursa or Beta Eridani, which lies close to Rigel in Orion. Today Kursa is unchanged, but Zibal is two magnitudes fainter—and that is a great deal.

If the fading is real, it must indicate evolutionary processes going on in the star, but Zibal does not seem to be that kind of object. It should be completely steady, and I suppose that we must simply distrust the old records, but it is certainly curious that they all made the same mistake. Moreover, there are other reported changes of secular variability; for example Castor, in Gemini, was once ranked as being brighter than its twin Pollux, but is now half a magnitude dimmer.

One final point. In general, only fairly bright stars have individual or proper names. Zeta Eridani, in its present state, would hardly merit a name of its own—and yet it has one. So was it really more prominent in earlier times? One cannot be sure.

C80 The Return of Mildred

I am delighted to be able to tell you that Mildred has been found at last. She has been missing for the past 75 years, and many people feared that she would never be seen again.

'Well', you may say, 'who is Mildred?' In fact this particular Mildred is not a person: she (or it) is an asteroid or minor planet, No. 878 in the list. The original discovery was made in September 1916 by two American astronomers, Harlow Shapley and Seth Nicholson, using what was then the world's largest effective telescope, the Mount Wilson 60-inch reflector. They named her Mildred after Shapley's baby daughter.

Asteroids keep mainly to the region of the Solar System between the orbits of Mars and Jupiter. They look like stars, and betray themselves only by their movements; only one (Vesta) is ever visible with the naked eye, and only one (Ceres) is as much as 500 miles in diameter. Mildred was a member of the main swarm, and was very faint, with a diameter of slightly less than two miles. The orbit was quite conventional, with a period of 3.63 years and a distance from the Sun ranging between 170,000,000 miles and 220,000,000 miles.

It is one thing to discover an asteroid, but quite another to work out its orbit well enough to make sure that it can be found again. After 1916, Mildred vanished. Very few numbered asteroids have been lost, and by 1986 there were only three absentees—719 Albert, 1179 Mally, and Mildred. Then Mally was recovered. The list of missing numbered asteroids dwindled to two.

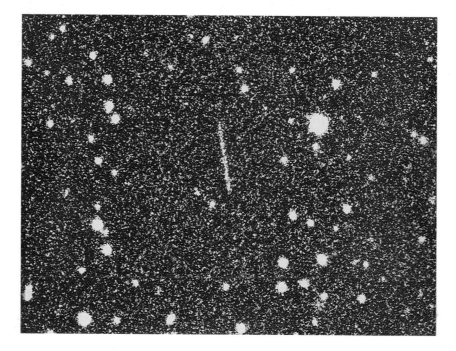

The recovery of Mildred, who shows up as a streak. Reproduced by kind permission of the European Southern Observatory.

On 10 April 1991 Oscar Pizarro, at the La Silla Observatory in Chile, took a photograph at the request of the Belgian astronomer Eric Elst, who was searching for asteroids of quite different type: Trojans, which move in the same orbit as Jupiter. When Elst received the plate he made a careful study of it, and found several asteroid images.

The next development was due to Gareth Williams, a young English astronomer who was interested in asteroids. In May 1991 he carried out some mathematical detective work, and concluded that one of the asteroids on the Pizarro plate might be Mildred. Checking back on earlier photographs, he was able to prove his point. Mildred had been photographed, though not identified, by L. V. Zhuravleva from the Crimean Astrophysical Observatory in 1985, three times from La Silla in 1977, and once from Siding Spring, in Australia, in 1984. It all fitted in, and the result was conclusive. Mildred was back—to the delight of Mildred Shapley Matthews, after whom the asteroid had been named, and who is now an editor at the Lunar and Planetary Laboratory in Arizona.

There is no fear that Mildred will be lost again, and that leaves only No. 719, Albert, last seen in 1911. Albert has an exceptional orbit which brings it close to the Earth, and is very small indeed. Whether it will ever be recovered now seems doubtful—but we thought the same about Mildred, so do not despair.

81 Exploding Star

Have you ever heard of R Aquarii? Probably not. It is not visible with the naked eye, and in general even binoculars will not show it, though on occasion it has been known to rise to the sixth magnitude—the limit of naked-eye visibility. Even when identified, telescopes show it as nothing more than a speck of light. For that matter Aquarius, the constellation in which it lies, is itself rather dim and formless, even though it is in the Zodiac and can sometimes contain planets. From Britain it is best seen during evenings in autumn.

R Aquarii, a mere 700 light-years away from us (so that we see it today as it used to be at the time of the Crusades) has interested astronomers for a long time. As long ago as 1811 the German observer Karl Harding discovered that it is variable, but although there was what seemed to be a very rough period of between 380 and 390 days the behaviour was often wildly erratic. One never knew quite what R Aquarii would do next. Then, in 1919, it was found that there is associated nebulosity, so that clearly the system was more complicated than it looked at first sight.

We now know that it is a symbiotic system, made up of two components. One is a very old star which has blown off its outer layers, so that all that is left is a small, very dense core; a star of this type is known as a white dwarf. Near it is a more normal star, reddish and cool but over 200 times as luminous as the Sun. This star periodically dumps material on to the white dwarf;

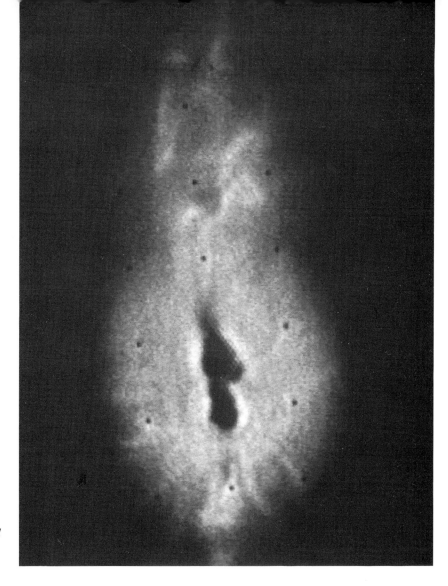

R Aquarii. The Hubble Space Telescope has provided this remarkable picture of the symbiotic star. Reproduced by kind permission of ESA and NASA.

material accumulates, and suddenly there is an atomic explosion which sends the outer layers of the star speeding away at the rate of several hundred thousand miles per hour.

Obviously it is difficult to study R Aquarii in any detail, but the Hubble Space Telescope has been of immense help. We all know about its faulty mirror, but it can still carry out excellent work, and its picture of R Aquarii is much the best ever taken. The two dark knots at the centre of the image seem to contain the binary system itself—the red star and the white dwarf. The knots are dark due to saturation effects produced by the camera in the Hubble Telescope (the Faint Object Camera: FOC for short) which are inevitable when the target bodies are relatively bright.

The picture also shows filamentary features coming from the core; these are made up of hot gas (plasma) which has been ejected at high velocity from the binary pair. The plasma emerges as a sort of geyser, at least 250,000 million miles long, which has been twisted by the force of the explosion and channelled upward and outward by strong magnetic fields; moreover, the flowing material seems to bend back on itself in a spiral pattern, due possibly to something getting in its way.

We do not know as much as we would like about the past history of R Aquarii, and we cannot be sure whether there has been one particularly violent outburst in the past. It is not unique; there are other symbiotic variables, of which the best-known is Z Andromedæ, but R Aquarii is the brightest of them. We are anxious to find out as much as we can, because heavy elements are built up in stellar explosions of this kind, and we are all made up largely of these elements. So what the Hubble Telescope is telling us about R Aquarii is highly significant.

☾82 Hagomoro

The Japanese have sent their first space-craft to the Moon. On 24 January 1990 they launched their Muses-A probe, which soared aloft from the Kagoshima Space Centre between the mountains and Japan's southern coast. It was sent up by a M3S-2 rocket, which is a slender three-stage affair, 92 feet long and no more than 5½ feet wide, with a weight of a mere 62 tons. It uses solid fuels, but on its own it would not have enough power to send a capsule directly to the Moon.

As was explained by Dr Hiroki Matsuo, director of the mission, the plan was to put Muses-A into a very elliptical path round the Earth, which would bring it within 11,000 miles of the Moon. Just before crossing the Moon's orbit it would release a smaller satellite with a 'kick' motor, and it would be this smaller craft which would go into a closed path round the Moon. Without carrying batteries, it had to remain in sunlight for the first thirty days, collecting enough solar power to send data back to the 'mother' satellite for transmission to Earth.

All went well, and on 24 January 1990 the lunar satellite, Hagomoro, was put into its closed orbit 10,000 miles above the Moon. The entire operation went exactly as planned. There was never any intention of landing Hagomoro on the lunar surface; its rôle was to collect data from orbit, not only about the Moon itself but about conditions in near-lunar space. Remember, that even though we have learned so much from the various missions—the Apollos above all—there is still a great deal about the Moon that we do not know.

Of course, this was not Japan's first foray into space. There have been plenty of artificial satellites, and there were the two probes to Halley's Comet in 1986, both of which functioned excellently and sent back a great deal of valuable information. But it is fair to say that the main purpose of the Hagomoro mission was to show that Japan was quite capable of joining in any lunar programme. Up to then all the expeditions to the Moon—manned or unmanned—had been either American or Russian, and the last of them, the Soviet Luna 24, dated back to August 1976. Luna 24 made a controlled landing in the waterless Mare Crisium (Sea of Crises), collected samples, and came home; since then the Moon has had no visitors from Earth, so that

Suisei space-craft.

the Japanese are opening a new era which may well culminate, before the end of the century, in a manned Lunar Base. To quote Professor Yasunori Matogawa, of Japan's Institute of Space and Astronautical Science, 'We have taken one step forward to planetary missions'.

It cannot be claimed that the venture was a complete success. Transmissions ceased abruptly on 19 March, and were never resumed, so that presumably there was an on-board fault which could not be corrected. However, by that time enough had been achieved to show that the overall outlook is encouraging.

Why Hagomoro? Well, this is the name of a feather robe which, in Japanese legend, was worn by a celestial nymph. In Japan the Moon has always had a mystical appeal, and this is still true today, but Hagomoro is not mystical; it is a very modern-type space-craft, and it means that the Japanese success with their probes to Halley's Comet was not an isolated triumph. All future space research must be truly international, and the Land of the Rising Sun is not likely to be left behind.

83 The Black Widow

In the little constellation of Sagitta, the Arrow, there is an object which is catalogued as PSR 1957+20. It is very faint, and looks innocent enough, but it is certainly unusual. We nickname it the Black Widow.

PSR 1957+20 is made up of two components. One of these is a dim star which is probably of the white dwarf variety, while the other is a pulsar—a stellar remnant, all that is left of a formerly very massive star which exploded as a supernova and sent most of its material away into space. The remnant is made up of neutrons, and is an incredible object. It is generally believed that a neutron star has a gaseous atmosphere, only a few feet deep, overlying a crystalline crust which is a thousand million million times as rigid as steel; what happens lower down is by no means certain, but the whole diameter of the star is only a few miles, even though the mass is as great as that of the Sun. The density at the core has been estimated at about a thousand million tons per cubic inch, so that a teaspoonful of it would weigh considerably more than a liner such as the *QE2*. The surface gravity is around 100 thousand million times stronger than that of the Earth. Visit a neutron star, and you will feel quite heavy. . . .

A neutron star has an immensely powerful magnetic field, and is spinning round rapidly. Beams of radiation are sent out from its magnetic poles, and as the star rotates these beams sweep across us, rather in the manner of a revolving lighthouse (though here we are dealing with radio waves, remember, not visible light). In some cases, as with the Crab Nebula pulsar, there is an identifiable optical source. The frequency of the pulses tells us how fast the neutron star is spinning round, usually at a rate of from rather less than one second up to several seconds.

PSR 1957+20 is different, because it is spinning round 622 times every second, and is one of a class called millisecond pulsars. We believe that long ago it became associated with another star—the present companion—and made up what is called a binary system, with the two components orbiting round their common centre of gravity. Then the massive star exploded as a supernova, leaving a pulsar. The pulsar's gravity began to pull a stream of material away from its companion, and as this stream fell toward the pulsar it 'speeded up' the pulsar's rotation (try blowing on one side of a table-tennis ball suspended on the end of a string, and you will see the effect). This is why the spin rate is now so tremendous.

But there is another factor to be taken into account. As the two bodies rotate, the pulsar-facing side of the companion is being intensely heated, and is literally being boiled away even though the distance between the two is of the order of a million miles. It has been calculated that in less than a thousand million years from now the companion will have been completely evaporated, and all that will be left of the system will be the murderous millisecond pulsar. It will have destroyed its mate, which is why we call it the Black Widow. We are watching a cosmical drama from a distance of 3000 light-years, and we know that even at the present time the doomed companion is left with a mass only about 25 times that of the planet Jupiter.

Another case has been found in the globular cluster Terzan 5, which is much further away from us. The Terzan pulsar, PSR 1744-24a, spins over 86 times in each second, but sometimes the radio pulses stop briefly, indicating that the pulsar has become engulfed in clouds of the material which it is wrenching away from its companion. There seems little doubt that we are dealing with another Black Widow; the universe is indeed a violent place.

84 Mini-Comets

In 1986, when Halley's Comet was causing so much general interest, a novel theory was proposed by Louis Frank, a physicist from the University of Iowa. He claimed that we are being bombarded regularly by millions of slushy 'mini-comets', which deposit water in our upper atmosphere.

Frank reached his conclusions from studying pictures of the upper air sent back by an artificial satellite, *Dynamic Explorer 1*, which was launched in 1981 and was put into a polar orbit, so that each revolution takes it over first one of the Earth's poles and then the other. Frank noticed strange dark spots in the pictures taken at ultra-violet wavelengths, and decided that these spots must be due to clouds of water vapour, each around 90 miles long. He reasoned from this that the spots were produced by slushy mini-comets, coming in to the Earth from all directions all the time.

We tend to think of comets as being icy, which is true enough; but by the time that Frank's mini-comets hit the upper air they are more like slushy snowballs. They could yield the equivalent of about 4/100 of a millimetre of rainfall over the Earth each year, and since the world is more than 4500 million years old this rate of deposition could account for all the water in our oceans. Larger comets may also have deposited the materials essential for life, such as carbon compounds, and this brings us back to the controversial ideas of Sir Fred Hoyle and Professor Chandra Wickramasinghe, who believe that life on Earth was first brought here by way of a comet.

Frank believes that his mini-comets could be detected by using ordinary telescopes fitted with highly sensitive light detectors. Each mini-comet would be around 30 feet in diameter, and would be moving at about 18½ miles per second at a height of from 3000 to 6000 miles above sea-level.

He has also discussed the possibility of mini-comets over the Moon. There is no trace of water there, but the immense pressure and high temperature created by a mini-comet's impact would not allow water to condense. Finally, he maintains that mini-comets represent the remnant of the disk of dust and gas from which the planets themselves were formed.

Scientists in general treated Frank's views with considerable reserve, but they have now been re-examined by John Olivero, of Pennsylvania University, with interesting results. While making observations of the varying amount of water vapour in the upper atmosphere, he found that there were sudden, temporary increases in the water vapour content every three days. He had not been concerned with mini-comets—they were not in his thoughts at all—but after 22,000 separate observations, each lasting for twenty minutes, he found that in over a hundred cases there was a genuine effect.

Olivero then looked back at Frank's data, and was impressed. Frank had calculated that in a region of the size which Olivero had observed there should be one mini-comet impact every two days, and this agreed fairly well with the three-day periodicity found by Olivero.

Despite this, most people are still sceptical. Reasonably large ice-balls of this kind would land unchecked on the airless Moon, but have never been observed. It has also been pointed out that any object of appreciable size would be detectable by modern ground-based radar, but there has been no sign of mini-comet activity. Finally, the sudden dumping of a large amount of water vapour would have marked effects on that region of the air known as the ionosphere, which reflects many radio waves, and again nothing of the sort has been observed.

We must, therefore, reserve judgement. Personally I am not convinced that we are being continually pelted by ice-balls, but one can never be sure.

C85   What is a Snu?

Have you ever heard of a Snu? I assure you that it is not a flightless bird, or a province in Nigeria. The letters stand for Solar Neutrino Unit.

A neutrino is a particle which is not easy to detect, because it has no mass and no electrical charge; also, it moves incredibly quickly. However, astronomers have had a certain amount of success in tracking them down, and we know that they come from space; in particular, they come from the Sun.

Remember, the Sun—like all normal stars—is simply a huge nuclear reactor. It is not burning in the conventional sense of the term, but is using nuclear

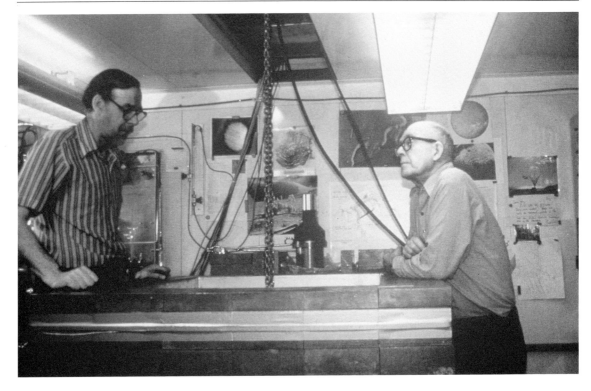

Inside the Homestake Mine Observatory: Ray Davies and Keith Rowley examine the equipment.

transformations taking place deep inside it, where the pressure is immense and the temperature is very high. Basically, nuclei of the lightest element, hydrogen, are banding together to form nuclei of the second lightest element, helium. Each time this happens, a little mass is lost and a little energy is released. The mass-loss amounts to 4,000,000 tons per second (please do not worry; there is plenty left) and the emitted energy keeps the Sun radiating.

During this process, which is admittedly somewhat complex, neutrinos are sent out. They flash across the 93,000,000-mile gap between the Earth and the Sun, and pass straight through us as though we did not exist. You are being bombarded with neutrinos at this moment, though it is quite impossible to notice them.

If a neutrino scores a direct hit on an atom of chlorine (one of the two elements making up common salt) it may change it into a special kind of the element argon, which is mildly radioactive, and which can be detected. However, the probability of this happening is not very high: about 10^{-36} per chlorine atom per second. 10^{-36} means 1 divided by 36 zeros . . . and this is one Snu.

The first serious experiments were started in the 1960s by Dr Ray Davies and his colleagues, of the Brookhaven National Laboratory in the United States. Davies filled a large tank with 100,000 gallons of cleaning fluid, which is rich in chlorine, and buried it a mile down in the tunnels of Homestake gold-mine in South Dakota; this was a necessary precaution, as otherwise there would have been interference from quite different particles known (misleadingly) as cosmic rays. Every few weeks he flushed out the tank to see how many chlorine atoms had been changed into radioactive argon. This, of course, would give the number of neutrino hits.

The results were surprising. There are about 10^{32} chlorine atoms in the tank, and there should be, in theory, about 7.9 hits per day. This was not what was found. The actual value was no more than 2.2 hits or Snus. Something was wrong—and it did not seem to be due to any fault in the apparatus.

Next came the setting-up of another detector, in the northern Caucasus Mountains: it was called Sage, which stands for Soviet–American Gallium Experiment. The principle was that when a neutrino hits an atom of the rare element gallium it may change it into an atom of a different element, germanium. Result—nil! And even allowing for experimental error, this is far too low a Snu. To make matters even worse, the Homestake Mine workers have now found that the Snu is lowest when the Sun is most active, near the peak of its eleven-year cycle.

One suggestion was that the core of the Sun might be rather cooler than had been believed. A value of 14,500,000 degrees Centigrade would help in solving the neutrino problem, but the generally accepted value was over 15,000,000 degrees, and reducing it would cause extra theoretical difficulties. Another idea was that the Sun might be behaving anomalously at the present time, but this would really be too much of a coincidence.

In 1990 there has been a new suggestion, due to three scientists: Stanislav Mikheyev and Alexei Smirnov in Russia, and Lincoln Wolfenstein in America. They point out that there are three kinds of neutrinos, some of which can be detected by the South Dakota and Sage equipment while others cannot. The new idea is that the different types can 'mix' with each other as they come from the Sun, so that detectable neutrinos are changed into undetectable ones. In this case the neutrinos are as plentiful as theory demands; it is just that we cannot record them.

If this is valid, astronomers will be immensely relieved. For years now they have been puzzled as to why the Sun seems to send out far fewer of these strange, elusive particles than it ought to do.

86 Yerkes Observatory: The Bees and The Crickets

The Yerkes Observatory, at Williams Bay not far from Chicago, is a remarkable place. Unlike most major observatories it is easy to reach; in fact it adjoins a golf course. Moreover its main telescope is not a reflector, but a refractor; the 40-inch lens is the largest in the world, and the telescope is still in use on every clear night. The Observatory itself was master-minded by George Ellery Hale, who was remarkably good at persuading friendly millionaires to finance his schemes; in this case the millionaire was Charles Tyson Yerkes, a Chicago street-car magnate. Yerkes insisted that the chosen site should be

The Yerkes 40-inch refractor, still the world's largest—and likely to remain so.

within a hundred miles of his home town, which is why Williams Bay was selected.

The building is classical, but there are two stone columns flanking the main doors which caused some discussion at an early stage. The design includes a bee which is about to sting a man on the nose. Just before the Observatory was about to be officially opened, one of the many trustees saw the bee designs, and was appalled. Not only was it undignified, but there was a lurking suspicion that the victim might be Yerkes himself, represented as about to be 'stung' for the money.

Action was promptly taken. A mason was called in, and solemnly chiselled away ninety-five bees. Look at the columns, and you can still see the scars where the bees used to be.

Another Yerkes story concerns the second Director of the Observatory, Edwin Frost: he succeeded Hale, who left after a few years to deal with the great new observatory which he was establishing at Mount Wilson, in California. For some reason or other Frost wanted to measure the temperature without having to use a thermometer—and the temperature range at Yerkes is considerable, with hot summers and very cold winters. Frost decided to make use of crickets, which are very plentiful in the area and are very much in evidence with their constant, friendly chirping.

This was Frost's method. Count the number of chirps made by a cricket in 13 seconds, add 40, and you have the temperature in degrees Fahrenheit. (Of course, if you want to convert to Centigrade you will have to do some mental arithmetic.)

I have never been able to test this myself, because every time I have been to Yerkes it has been the close season for crickets, but I am told that it works. If you doubt me, try it for yourself next time the local crickets are in full cry.

87 Flying for Halley

1986 was the year of Halley's Comet. Purely as a spectacle, the comet was a failure; it was not nearly so bright as it had been in 1910 or in 1835. Nevertheless, the interest which it aroused was tremendous, and everybody seemed to want to see it. Things in England were not helped by the fact that from November through to mid-January the weather was atrocious; we usually have a good deal of cloud and rain, but this time it was much worse than usual. On the few clear nights I did my best, and I showed the comet to as many people as possible. I suppose that from four hundred to five hundred viewers came through my observatory in one week alone.

Yet it was quite true that many people who wanted to have a good look at the comet were unable to do so, because of the cloud. Then I had a call from British Airways. Would I go up with a plane-load of passengers and show them the comet from an altitude of 35,000 feet, well above the cloud-tops?

It sounded rather fun, and from terra firma I could not see the comet anyway, so I said 'yes'. In the end I did half a dozen flights, and it was a strange experience.

The first problem is, of course, the aircraft window, even in a Jumbo Jet. Viewing a faint object such as Halley's Comet through a thick pane of glass is not easy at the best of times, and I had to break the news to one earnest passenger that it would not be at all a good idea to open the window. Few people knew their star patterns, and so it was helpful when Halley was in the same area as a bright star which I could point out easily, but this did not happen very often. For the first few flights the comet was reasonably high up, which meant that it was above the top of the aircraft window unless the pilot tipped the plane at a crazy angle. Admittedly this improved for the later flights, but it was replaced by another hazard—the Moon, which became more and more obtrusive, finally drowning the comet altogether so far as

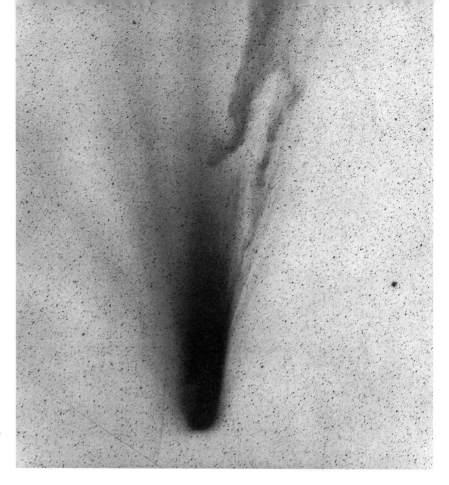

Halley's Comet, 10 March 1986, as photographed with the 1-metre Schmidt telescope at La Silla. Reproduced by kind permission of the European Southern Observatory.

naked-eye observers were concerned. In fact, it had been necessary to use binoculars for most of the time in any case.

Also, there were many people who expected to see a brilliant comet, with a long tail, streaking across the sky. I explained over and over again that this was simply not so, but I suppose that I did not sound convincing.

It was also true that the only person with a hope of seeing anything at all was the occupant of the window seat. So when he (or she) had managed to glimpse the comet, I had to persuade him to change places with someone who hadn't. Generally there was no real problem over this, but naturally there were some people who were quite incapable of seeing the comet or anything else, and remained firmly in their window seats throughout. We had one plane-load of teenagers, and I must say that they were the best-mannered of all, as well as being the most knowledgeable.

Of course there were the eccentrics. One astrological enthusiast was convinced that the end of the world was at hand, and that Halley's Comet was responsible; he wanted to have one last view of the cosmos before it was snuffed out. There was also the lady who was terrified of climbing too high in case we actually collided with the comet. I did my best to explain that we were going up for less than half a dozen miles and that the comet was some 50,000,000 miles away, but it was hard to reassure her. Someone else wanted to know if we intended to fly right outside the Earth's atmosphere; I explained that even a Jumbo Jet would find that a little difficult.

In the end most people did see the comet, and all told we took about a thousand people on our flights. They saw nothing but a fuzzy patch, which is all that Halley could produce, but it was better than nothing. One or two of the youngest passengers may well see it next time round, in 2061, and there was one old lady who had seen it in 1910 and was overjoyed to make its acquaintance once more. So I feel that these somewhat crazy flights were worth while.

☾88 What is Happening to our Pole Star?

One of the few advantages which northern-hemisphere observers have over dwellers in Australia or South Africa is that we have a much better Pole Star. The north pole of the sky is marked within one degree by a star of the second magnitude, Polaris in Ursa Minor (the Little Bear). By comparison the

Wandering of the Pole. Polaris has not always been the pole star; the position of the celestial role described a circle in the sky in a period of about 25,000 years. In Egyptian times the pole star was Thuban, in Draco (the Dragon). Today it is Polaris. In 12,000 years' time the nearest bright star to the pole will be Vega, in Lyra.

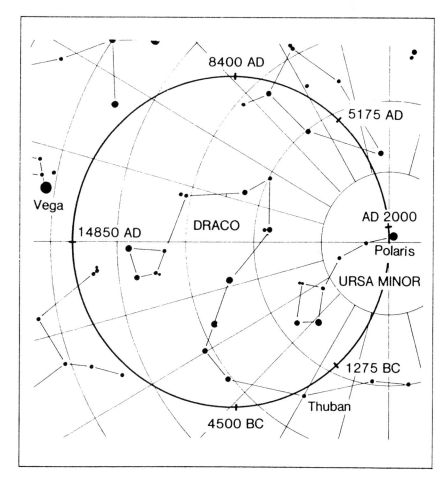

southern pole star, Sigma Octantis, is so dim that it is none too easy to see with the naked eye.

Polaris is easy to find, mainly because it seems to remain almost motionless in the sky with everything else revolving round it once in 24 hours. In case of any doubt, use the Pointers in Ursa Major, the Great Bear or Plough, which show the way to it. Neither of the Bears ever sets over the British Isles, so that they can always be found whenever the sky is sufficiently dark and clear. Apart from Polaris and one other star, Kocab, Ursa Minor is decidedly obscure, but once identified it will not be forgotten.

Polaris is powerful. It is around 6000 times as luminous as the Sun, and is 680 light-years away, so that we now see it as it used to be at the time of the Crusades. Its spectral type is F8, which means that its surface is slightly hotter than that of the Sun. In theory it should be slightly yellowish, but I have never been able to see any colour in it.

For a long time we have known that Polaris is not absolutely steady in its light, but fluctuates slightly between apparent magnitude 1.92 and 2.07, giving a mean value of 1.99; the period is 3.969778 days. Polaris is what is known as a Cepheid variable, and pulsates, swelling and shrinking regularly and changing its output as it does so. The changes in brightness are so slight that they cannot be noticed with the naked eye, but they are real enough, and it seems that most stars go through a stage of this type during their life-stories. It has been estimated that the Pole Star has been pulsating for 40,000 years at least.

In 1987 Canadian astronomers, using very sensitive measuring equipment on a 16-inch telescope in British Columbia, found that something strange was happening. According to Nadine Dinshaw, one member of the team, Polaris was not pulsating nearly as much as it had done a few years earlier. Using a more powerful telescope—the 47-inch reflector at the Dominion Astrophysical Observatory—this was confirmed; the Pole Star was indeed steadying down. Looking back in the records, it now seems that the change began in the early 1970s.

It is too early to say whether the Pole Star will stop pulsating altogether in the foreseeable future. It may well do so, and in this case we will actually have caught a star in the process of evolving from one stage to another. Most things in the universe happen very slowly, apart from nova and supernova outbursts, so we may have been fortunate.

Polaris has always been regarded as a very special star, and navigators have a great affection for it, but this has been simply because of its unique position in our sky. Now, however, astronomers are paying attention to it for quite different reasons.

Sudden Death

Astronomers in general tend to be long-lived, but there have been some who have been cut off in their prime. One of these was Jeremiah Horrocks, born near Liverpool in 1619. He studied at Cambridge for the Anglican ministry,

Karl Schwarzschild. Brilliant German astronomer—and war victim.

but his main interest was in astronomy, and he was the first to observe a transit of Venus, in 1639. Venus does not often pass across the face of the Sun—the last occasion was in 1882, the next will be in 2004—but Horrocks had made his own calculations, and he was clearly a mathematician of exceptional promise. Unfortunately he died at the age of twenty-one.

Horrocks' death was due to natural causes (at least, so far as we know), but this was not the case with his friend William Gascoigne, who was one year younger and came from Middleton in Yorkshire. By the age of eighteen he was already making useful observations, but his real claim to fame is that he invented the micrometer, a special instrument which is used together with a telescope to measure very small angles and distances, such as those between the components of a double star.

Gascoigne's first micrometer seems to have been made in or about 1638. He used it well, making accurate measurements of star positions and the apparent diameters of the Sun and Moon. He also worked out improvements for telescopic lenses, and he seemed set for a brilliant career. Sadly, he was given no chance. In 1643 the civil war between Charles I's Cavaliers and Oliver Cromwell's Roundheads broke out, and Gascoigne joined the Royalist forces.

What is often regarded as the turning-point of the war was the Battle of Marston Moor, fought seven miles west of York on 2 July 1644. The Cavaliers, led by Prince Rupert of the Rhine, were heavily outnumbered, but in the end Cromwell's cavalry was decisive. Rupert's army lost four thousand men— and one of these was William Gascoigne. Apparently he had a major work on optics ready for the press, but after his death no trace of it could be found.

Another war casualty, much later, was the great German astronomer Karl Schwarzschild, who was born in 1873. He went to the Vienna Observatory, and then to Göttingen, where he carried out important work in astrophysics. For example, he was the first to define what we now call the colour index of a star. Photographic plates of those days were more sensitive to blue light than to red, so that if a red star and a blue star appeared equally bright to the eye they would leave different impressions on the plate—the blue star would seem much the more prominent, and red stars such as Betelgeux in Orion would make a very poor showing. Colour index is the difference between the blue and the red images—the greater the colour index, the redder the star; the scale was adjusted so that a pure white star, such as Sirius, would have a colour index of zero.

Schwarzschild measured the colour indices of over 3500 stars, and when he went to Potsdam in 1909, as Director of the Observatory, he concentrated also upon stellar movements and the structure of the Milky Way. Then, when he was forty years old, war broke out. He volunteered for the German Army, and after service in Belgium and France was sent to the Russian front. It was here that he contracted a skin disease for which there was no cure. Despite this, he continued with his mathematical work, and it was during the last part of his life that he worked out the equations upon which we now base most of our ideas about those bizarre objects which we call black holes.

Schwarzschild examined the case of a spherical body which was very small and very massive. If it shrunk below a certain critical radius, he claimed,

Gascoigne's Leap—a plaque in the dome of the Anglo-Australian Telescope!

it would distort space-time so severely that nothing—not even light—could escape from it; it would become a black hole. For a body the mass of our Sun, the Schwarzschild radius is a mere 1.75 miles.

Schwarzschild had just time to finish these calculations before he died, in 1916. His official obituary was written by Albert Einstein.

World War 2 claimed the life of a Russian, Leonid Kulik, who was one of the foremost experts on the subject of meteorites. In 1927 he had led the first party to reach the site of the great explosion which had taken place on 30 June 1908 in the Tunguska region of Siberia, when an object had fallen from the sky and blown pine-trees flat over a wide area. In July 1941, when Hitler invaded the Soviet Union, Kulik joined the Moscow People's Militia; in October he was wounded in the leg, captured by the Germans and taken to a Nazi prison camp, where he died in April 1942.

There have also been accidents. Ernst Klinkerfues, Director of the Göttingen Observatory, was killed in 1884 when he fell from his observing platform while studying a comet, and much more recently—in 1987—there was the tragic death of Marc Aaronson, a young, brilliant and very popular American astronomer, who was crushed by a door at the Kitt Peak Observatory during what should have been a routine night's work.

One man who had a lucky escape was Dr Gascoigne (no relation to the inventor of the micrometer) who was walking along the observing platform at the Siding Spring Observatory, in New South Wales, when he overlooked a gap in the rails and plunged through, falling at least twenty feet. Mercifully he did no more than sprain an ankle, but the site is now marked by a small plaque with the inscription: *Gascoigne's Leap!*

☾90 The Two-faced Moon

I always think that Saturn is the loveliest object in the whole of the sky. When the rings are wide open as seen from Earth—as they are during the early 1990s—they are magnificent. They may look solid, but they are not: they are made up of vast numbers of small icy particles, all speeding round the planet in the manner of miniature moons.

Moreover, Saturn has an extensive family of true satellites. Eighteen are known, and a few more have been suspected. One of them—Titan—is much larger than our Moon, and has a dense atmosphere made up largely of nitrogen. Next in size come Rhea, Iapetus, Dione and Tethys, all of which are bright enough to be seen with a modest telescope. The rest are smaller, and the junior members of the family are known only from the images sent back from the two Voyager space-craft.

Iapetus, discovered by G. D. Cassini as long ago as 1671, is of special interest. It is two-faced; one side of it is almost as bright as snow, while the other side is blacker than a blackboard. This has been known for a long time. Iapetus takes 79 days to complete one journey round Saturn, travelling in a fairly circular orbit at a mean distance of 2,200,000 miles from the centre of the

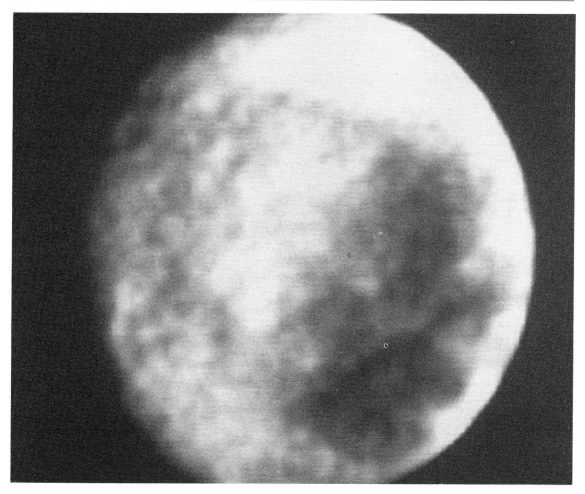

Iapetus, as photographed by Voyager 2.

planet. When seen to the west of Saturn it is conspicuous, with a magnitude of between 9 and 10 according to my estimates; when east of Saturn it is much fainter, and drops down to below magnitude 12. There is no mystery about this. Like most major satellites, it has a captured or synchronous rotation—that is to say, its orbital period is equal to its axial rotation period—so that it always keeps the same face turned toward Saturn. It follows that when we see it lying to the west of the planet, we are seeing its brighter hemisphere.

Thanks to the Voyagers (particularly No. 2) we have a reasonable map of much of Iapetus, and we can make out the usual hilly, cratered terrain. It is the 'leading' hemisphere which is dark; the line of demarcation is not abrupt, and there is a sort of transition zone between 100 and 200 miles broad, while the boundary itself is rather irregular. One very curious feature is a ring of dark material close to the boundary, about 250 miles in diameter. Its nature is unknown; it does not seem to be a crater, and it is unlike anything else seen on Iapetus or the other Saturnian satellites. There is a wide bright plain, named Cassini Regio in honour of the Italian astronomer who discovered Iapetus as well as three of the other satellites (Rhea, Tethys and Dione).

Unfortunately, we do not know much about the dark, following hemisphere, though there is no reason to doubt that it too is cratered.

It is not easy to see why Iapetus, and only Iapetus, is two-faced. We have to deal first with what I call the zebra problem: is a zebra a black animal with white stripes, or a white animal with black stripes? Fortunately we can find out the mass and density of Iapetus, from its effects on the other satellites. The diameter has been given as 892 miles; the density is only 1.2 times that of water, so that the globe must be made up of rock together with a great deal of ice. This means that it is basically bright, with a black surface stain.

Spectral analysis has shown that the bright part is covered with ordinary water ice, while the blackish regions are coated with carbon compounds. It has now been found, by E. A. Cloutis in Canada, that the dark surface resembles terrestrial tar sand—a combination of clays, quartz grains, hydrocarbons and water, with traces of other minerals.

Well beyond Iapetus, at over 8,000,000 miles from Saturn, moves the outermost satellite, Phœbe, which is small, dark and moves in a retrograde or wrong-way direction, so that it may be a captured asteroid rather than a bona-fide satellite. There have been suggestions that dark material has been swept off Phœbe on to Iapetus, but the colour is wrong—the coating on Iapetus is too reddish—and in any case Phœbe is too small and too far away; the two satellites are always more than four million miles away from each other. Another idea is that the dark material has welled up from inside the globe, in which case it could be a mixture of ammonia, soft ice and a dark substance of unknown origin. We do not know how deep the layer is; it could amount to only an inch or two, though it could also be quite thick. The latest suggestion, due to R. G. Tabak and W. M. Young in America, is that at some time during the past hundred thousand years Iapetus was hit by a comet. The volatile ices dissipated, but the darker material re-accumulated on the leading hemisphere over the next million years or so.

No doubt we will find out eventually. There is no hope of sending expeditions there in the near future, but the first travellers to land on Iapetus will have a grand view of Saturn, because the orbital inclination is almost 15 degrees, whereas all the inner satellites move practically in the plane of the planet's equator, and from them the rings will always appear edgewise-on. There is, of course, no atmosphere—the escape velocity is less than half a mile per second—and so the sky will be jet-black. So far as we know at present, the zebra-like Iapetus is unique.

C91 Space Débris

Space Shuttle, Discovery 1991 as I photographed it before launch from Cape Canaveral.

On 16 September 1991 the astronauts on the US Shuttle *Discovery* were faced with what could well have been a very dangerous situation. They were travelling at 352 miles above ground level, moving at a speed of 17,500 m.p.h., when Norad, the North American Air Defence System, found that a piece of 'space junk' was on a near-collision course. The item was in fact part of an abandoned Soviet rocket, and was identified as the upper stage of the vehicle which had launched Cosmos 955 as long ago as September 1977. It was moving at 15,600 m.p.h., and it seemed likely to pass within 1.7 miles of the Shuttle. If it had impacted, the relative velocity would have been over 2,000 m.p.h., and since the débris was about the size of a minibus the result would have been total disaster. No space-craft could have survived such a blow.

Luckily there had been five hours' warning, and *Discovery*'s orbit was promptly lowered by ten miles, so that the launcher by-passed it at a safe distance. The flight director, Al Pennington, commented that the situation had been 'uncomfortable', and added judiciously that 'We don't like to take risks'. Incidentally, the actual launch of *Discovery* had been delayed for a few minutes to allow known orbiting débris to pass by.

This episode highlights a problem which is increasing all the time. Larger objects in space can be followed, but there is also a great deal of rubbish which cannot possibly be detected. NASA has calculated that there are at least seven hundred orbiting objects which are potential hazards, ranging from pieces of rockets and satellites to discarded tools, pieces of rag, and even an astronaut's glove. Just as harmful are even smaller objects, because when moving at these tremendous velocities pieces of material no larger than pins' heads can be transformed into lethal missiles. One space-craft screen was indeed damaged by a hit from an object which seems to have been nothing more nor less than a fleck of paint.

Much of the problem was created in the early days of space research, when rockets and satellites were less reliable than they are now. Several satellites actually exploded, and put hundreds of fragments into orbit. While some of them have been braked by friction against the air, and have burned away in the lower atmosphere, others have not, and they remain as a growing menace.

Great attention is now being paid to the problem. What is needed, of course, is a cleaning-up operation, but just how this can be achieved is not clear; at best it will cost a vast amount of money, and will need full international co-operation. It would probably be possible to remove objects of football size or larger, but the rest will presumably have to stay in orbit until they come down of their own accord—even the astronaut's glove.

Certainly there is need for prompt action. Otherwise, it is quite likely that within the next few decades travelling in low orbits round the Earth will become unacceptably dangerous.

92   Geminga Traced!

Bodies in the sky send out radiations at all wavelengths, from the very long (radio waves) through microwaves, infra-red, visible light, ultra-violet, and X-rays through to the very short gamma-rays, which cannot be studied from down here because they are blocked out by the Earth's air. For gamma-ray astronomy, space research methods are essential.

One very strong gamma-ray source has caused a tremendous amount of discussion. It lies in the constellation of Gemini, the Twins, and is known as Geminga. It was identified by an artificial satellite launched as long ago as 1972, and proved to have a gamma-ray spectrum not unlike that of another powerful source which is situated in the southern constellation of Vela, the Sails. But the Vela source was known to be a pulsar, and could be traced both because of its radio signals and its optical flashing. Geminga was much more of a problem. It was thought to be a pulsar, but it sent out no detectable radio waves, and neither was there any visible source in its position.

Astronomers kept on looking for Geminga—or, to give it its catalogue title, 2CG 195+04. At last X-rays were detected from about the same position in the sky, which appeared to be promising. Then, in 1983, came the launch of a new satellite, the Einstein Observatory, which was designed specially for X-ray work. Using it, an Italian team led by Giovanni Bignami set out on a new hunt for Geminga. They confirmed that X-rays came from the expected position, but what about a visible star? One, of magnitude 21, was an initial candidate, but closer examination showed that it was nothing more than a foreground star of much the same type as the Sun. Geminga had won the second round.

Close by, however, were two much fainter stars, catalogued as G1 and G2. They were so faint—magnitudes 24.6 and 25.5 respectively—that optically they were very near the limit of detection. In 1986 two American astronomers, Jules Halpern and David Tytler, decided to make a new survey with the help of the 200-inch Palomar reflector, which was for so long much the most powerful telescope in the world. They found that G2 was unusually blue, and this, together with its X-ray spectrum, indicated that it could be an isolated neutron star from 1500 to 3000 light-years away, and with a surface temperature of between 500,000 and 1,200,000 degrees Centigrade. If so, then it is almost certainly Geminga. We receive no radio pulses from it because the beams coming out from the magnetic poles of the neutron star do not point in our direction.

It looks, therefore, as though we have solved the problem. Geminga has baffled us for many years, but at last it has had to tell us where it is.

93 Worlds of Curious Shapes

The asteroids or minor planets have not always been the most popular members of the Solar System so far as astronomers are concerned. In former times, photographic plates taken for quite different reasons were often found to be crawling with asteroid trails, all of which had to be eliminated and all of which wasted an incredible amount of time. One infuriated German observer went so far as to dub them 'vermin of the skies'.

We think differently now, and we regard the asteroids as both interesting and important. They have diverse characteristics, and, for example, some of them are very irregular in shape. One would expect a large body to be

Asteroid Eros.
Photograph by Paul
Doherty.

spherical, and this is certainly true of Ceres, the senior member of the swarm, but Asteroid No. 2, Pallas, is not; it is triaxial, and measures $360 \times 330 \times 292$ miles. No. 3, Juno, is elliptical, with a longer axis of 179 miles and a shorter axis of only 143 miles.

Some asteroids are suspected of being double. One is No. 624, Hektor, which moves in the same orbit as Jupiter, well beyond the main swarm (though it keeps prudently well away from the Giant Planet, and is in no danger of being swallowed up). The diameter is of the order of 150 miles, which by asteroidal standards is large. Another exceptional body is 288 Glauke, no more than 20 miles across, which is distinguished its very slow rotation; variations in its brightness indicate that it takes 1500 hours to spin once round—over two months. If this is correct, the 'day' on Glauke is very long indeed. But most asteroids are much less leisurely; one of them, 1566 Icarus, has a rotation period of only 2 hours 16 minutes, though it is admittedly very small (and is one of only two asteroids known to approach the Sun closer than the orbit of Mercury; the other is Phæthon). So it may well be that Glauke is made up of two bodies moving around each other rather than a single asteroid spinning very slowly.

The first asteroid known to move within the orbit of Mars was discovered in 1898; it is No. 433 in the list, and was given the name of Eros (all previous asteroid names had been female). Eros is shaped like a sausage, less than 20 miles long. This sort of thing is not unexpected; after all, asteroids are cosmical débris, and there have no doubt been frequent collisions. No large planet could form in that region of the Solar System, because of the tidal disruption caused by massive Jupiter.

When we consider the very junior members we find some strange things. The very smallest asteroid so far identified has not yet been named, and is referred to only as 1991 BA. In January of that year it brushed past us at only 106,000 miles, which is less than half the distance of the Moon, but it was still very faint, because it is so tiny. The estimated diameter is 30 feet. You could not mark out a cricket pitch on 1991 BA; there would not be enough room on the whole of the asteroid!

Another small body, 1989 PB, by-passed us at 2,500,000 miles, which is close enough for it to be contacted by radar. Radar involves sending out a pulse of energy and bouncing it off a solid body or equivalent; the reflected pulse can be picked up, and can tell us a great deal about the target object. S. J. Ostro and his colleagues used the radio telescope at Arecibo, in Puerto Rico, to contact 1989 PB, and found to their surprise that it is shaped like a dumbbell. There are two distinct bells, each about half a mile in diameter, either touching each other or else joined together.

It will be fascinating to see what we will learn when we are able to send space-craft to obtain close-range pictures of these little worlds. Undoubtedly we are going to find some bizarre shapes among the asteroids!

☾94 Browsing

Browsing through old books is always a fascinating pastime; one is always coming across something unexpected. For example, is it possible for a comet to be inhabited? Johann Heinrich Lambert certainly thought so, as we learn from his *Lettres Cosmologiques*, published in 1865:

> 'The comet of 1680, being 160 times nearer to the Sun than we are ourselves, must have been subjected to a degree of heat 25,600 times as great as we are. Whether this comet was of a more compact substance than our globe, or was protected in some other way, it made its perihelion passage in safety, and we may suppose that all its inhabitants also passed safely. No doubt they would have to be of a more vigorous temperament and of a constitution very different from our own. But why should all living beings necessarily be constituted like ourselves? Is it not infinitely more probable that amongst the different globes of the universe a variety of organizations exist, adapted to the wants of the people who inhabit them, and fitting them for the places in which they dwell, and the temperatures to which they will be subjected?'

Lambert believed that a comet was a solid, rather planet-like body enveloped in a dense atmosphere, but a different tack was followed by Margaret Bryan, who published a book about general astronomy in 1799. Here is what she has to say about comets:

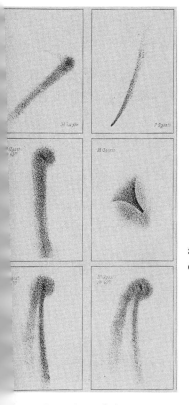

Drawings of the comet Swift-Tuttle, made by A. Secchi. This was in 1862; the comet has not been seen since which is why there are no photographs of it.

Saturn's Rings. On 12 November 1980 Voyager 1 took this photograph, from a range of 446,000 miles; the complex nature of the ring system is very evident. Reproduced by kind permission of NASA.

'These extraordinary bodies are found, by their reflective power, to be opake.* The matter of heat and light darts from them like fiery tails;—as when an insulated jar is receiving a full charge from the electrical machine, it throws off its redundancy, so do the comets emit a stream of fire from their bodies on the side opposed to the Sun, from which they receive their superabundant fire;—therefore, if I may be allowed to reason from analogy, I should suppose that by this mode do the comets throw off the redundant heat which they must receive from the Sun. . . . I venture to intimate the possibility of this being the cause of the effect we perceive of these motions called fiery tails;—that the additional heat thus issuing from these bodies prevents an accumulation unfavourable to animal existence, supposing these bodies to be inhabited, as we have great reason to do, seeing that God has created nothing in vain.'

According to Pierre Louis de Maupertuis, who lived from 1698 to 1759 and was a noted astronomer and mathematician (and, incidentally, a Fellow of the Royal Society), a comet might even come upon a raiding expedition:

'Not only might a comet carry away our Moon, but it might itself become our satellite, and be condemned to perform its revolutions about our Earth and illuminate our nights. Our Moon might originally have been a small comet which, in consequence of having too nearly approached the Earth, has been made captive of it. Jupiter and Saturn, bodies much larger than the Earth, and whose pull extends to a greater distance, and over larger comets, would be more liable than the Earth to make such acquisitions; consequently Jupiter has four moons revolving about him, and Saturn five.'

François Arago, the leading French astronomer of the first half of the nineteenth century, went so far as to consider what would happen if a comet captured the entire Earth, and turned it into a cometary satellite. When the greedy comet approached perihelion,

'There is no doubt that at first the solid envelope of the Earth would experience a degree of heat 28,000 times greater than that of summer; but soon the seas would turn into vapour, and the thick beds of clouds arising therefrom would protect it from the conflagration, which at first would seem to be inevitable. . . . At aphelion, the heat received from the Sun by the Earth would be 19,000 times less than the mean heat at present. Concentrated in the focus of the largest lens, it would certainly produce no sensible effect even upon an air thermometer. The temperature of our globe would then depend only upon the heat which might remain undissipated of that which it had received during its perihelion passage.'

However, he comes to the surprising conclusion that even if the Earth were taken in tow by a comet, 'there is nothing to prove that the human race would disappear through the effects of temperature.'

*Her spelling.—PM.

At least he admitted that the probability of an event of this kind is 'extremely small'.

What, then, about meteors? It was still generally thought that shooting-stars were atmospheric rather than astronomical phenomena, and in Mrs Byran's book a meteor is described as an 'appearance in the sky of a transitory nature, such as clouds, thunder, & c.' But it is when dealing with the aurora that she really comes into her own:

> 'The Aurora Borealis has been by some ascribed to a subtile effluvia, which, entering at one pole, is ejected at the other with violence, in order to restore an equilibrium. . . . Various other hypotheses have been formed, but nothing certain is ever likely to be understood of this extraordinary and partial illumination.—We may amuse ourselves as much as we please on this subject, but we must be satisfied with conjecture and the greatest stretch of our sagacity.'

Finally, let us look back to the views of a famous last-century astronomer and theologian, the Rev. Dr Thomas Dick, in his book *Celestial Scenery*, published in 1837. His views about the planet Saturn hardly agree with the information sent back by the Voyagers—but Dr Dick was nothing if not positive in his outlook:

> 'This planet is about 79,000 miles in diameter, and nearly a thousand times larger than the Earth. Its surface contains more than 19,600 millions of square miles, and, consequently, at the rate of 280 inhabitants to a square mile, it would contain a population of 5,488,000,000,000, or about five billions and a half, which is six thousand eight hundred and sixty times the present number of inhabitants on our globe; so that this globe, which appears only like a dim speck on our nocturnal sky, may be considered as equal to six thousand worlds like ours; and since such a noble apparatus of rings and moons is provided for the accommodation and contemplation of intelligent beings, we cannot doubt that it is replenished with ten thousand times ten thousands of sensitive and rational inhabitants; and that the scenes and transactions connected with that distant world may far surpass in grandeur whatever has occurred on the theatre of our globe.'

No comment!

95   The Geysers of Triton

When I look back to the time when I first became interested in astronomy, over sixty years ago now, I find it hard to realize how little we knew about some of our neighbour worlds. For example, we were confident that there were extensive vegetation tracts on Mars, while Venus was generally taken

Full-scale model of Voyager 2. I am standing by it to show its size. The model is on display at the Jet Propulsion Laboratory in California.

to be a friendly world, with primitive life-forms surviving quite happily in its warm seas. About the outer planets, Uranus and Neptune, our ignorance was fairly complete, and Pluto had not even been discovered.

However, we did at least know that Neptune had a large satellite, Triton; it had been discovered by William Lassell, a brewer-turned-astronomer, not long after Neptune itself had been found in 1846. It was unusual inasmuch as it orbited Neptune in a wrong-way or retrograde direction; its diameter was thought to be greater than that of our Moon, and perhaps as large as 6000 miles. There were also suggestions that it might have an appreciable atmosphere, but this atmosphere could not be like ours, simply because Triton was too cold. The mean distance from the Sun is 2,793,000,000 miles, and there is very little sunlight. Nitrogen, which makes up 78 per cent of the air we breathe, would presumably turn into a liquid.

So were there nitrogen oceans on Triton? It seemed a possibility. When Voyager 2, launched in 1977, drew in toward the Neptunian system in 1989 it was thought that Triton might be the pièce de resistance of the entire mission,

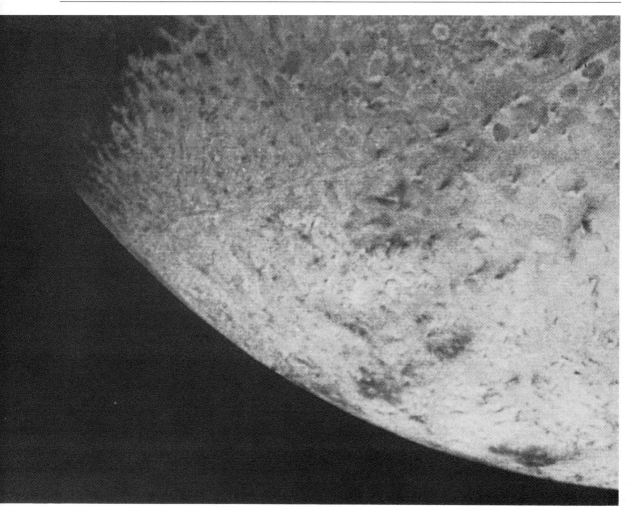

Triton, from Voyager 2; the dark streaks mark the geysers. Reproduced by kind permission of NASA.

and I recall one graphic prediction made two years before the space-craft was due to arrive: 'Perhaps Voyager 2, turning its cameras on Triton, will see plains of white and coloured organic deposits and, maybe, the glint of a distant sun reflected off a calm nitrogen sea.'

Triton did indeed prove to be a revelation, but not quite in the way that had been anticipated. First, it was smaller than our Moon, not larger; the diameter is only 1681 miles, as against 2158 miles for the Moon. It is also colder with a temperature of −233 degrees Centigrade. Its excessively thin atmosphere is composed of nitrogen, and its south pole is covered with pink snow—not water snow, but nitrogen snow.

The main shock came when Voyager 2 sent back close-range pictures of dark streaks in the pinkness which have been found to be active geysers, sending up plumes of material to at least five miles above Triton's surface. Activity in so chilly a landscape was the last thing that anyone had expected, but the evidence was conclusive, and some explanation had to be found.

Photomosaic of Triton: Voyager 2, assembled from 14 frames. The remnant of the S. polar cap is shown at the bottom of the picture, and the dark streaks of the geysers are evident. Reproduced by kind permission of NASA.

According to Dr Laurence Soderblom, one of NASA's leading planetary geologists, the surface of Triton is covered with a layer of nitrogen ice about three feet deep. It moves round with the Tritonian seasons, which are very long in view of the fact that Neptune takes 165 Earth years to complete one orbit. Underneath the ice layer we come to dark organic polymers, which are associations of molecules. When sunlight penetrates through, it warms the dark material. A rise in temperature of only four degrees increases the pressure underneath the ice by a factor of 10. Then a weak point in the ice leads to an explosion, and the result is a geyser. Each plume seems to be about 60 feet in diameter, sending out gas and dust at 5 kilogrammes per second. Over a period equivalent to one Earth year, several cubic feet of nitrogen ice are vaporized.

Alternatively, it has been suggested that the geysers are similar in origin to 'dust devils', atmospheric vortices which are localized but violent. Patches of unfrosted ground near that point on Triton's surface which is directly below

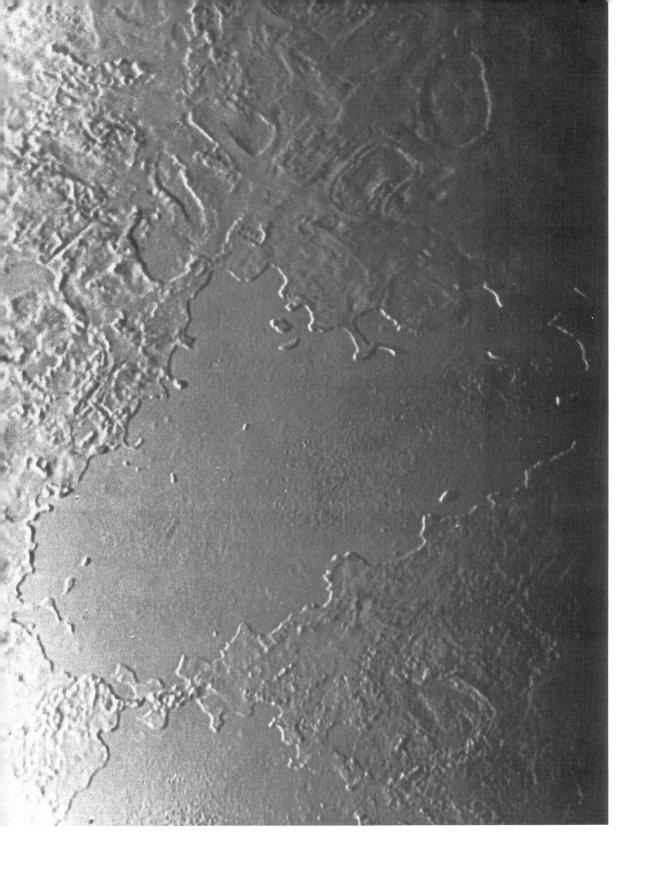

the Sun could heat up relative to the surroundings, and produce dust devils at the required rate of up to 60 feet per second.

All in all, the first of these theories seems to be the more plausible, but we cannot be sure, and there will be no further close-range pictures until a new Neptune probe is dispatched. Triton is at least a very curious place, and deserves to be studied, even though I doubt whether any astronaut will feel inclined to visit a world with a temperature of -233 degrees, pink snow, and whirling or explosive nitrogen geysers.

96 Dusty Space

For a long time we have known that the universe is a dusty place. In the huge systems of stars which we call galaxies there is a tremendous amount of dust, and it hides some of the stars. We also know of clouds of material known as nebulæ, which are made up of dust together with gas; it is in these nebulæ that new stars are being produced—and perhaps five thousand million years ago our Sun was born in precisely this way.

A new suggestion, by the Dutch astronomer Edwin Valentijn, is that there may be even more dust than we had believed. He has been studying photographs of external galaxies, many of them taken by the telescopes at the European Southern Observatory station at La Silla in Chile, and he has come to some rather unexpected conclusions.

Galaxies are of various shapes and sizes. Some, such as our own Milky Way, are spiral; so is the Andromeda galaxy, but the narrow tilt as seen from Earth rather spoils the effect. On the other hand the Whirlpool Galaxy, Messier 51 in the constellation of the Hunting Dogs, is almost face-on to us, and gives the impression of a majestic Catherine-wheel. If a spiral galaxy is edgewise-on to us, of course, we will see it in a different guise, and it may not be easy to make sure that it is spiral at all.

Valentijn claims that we see far fewer edgewise-on galaxies than we ought to do if they were oriented at random (as seems almost certain). Therefore it follows that there must be a vast amount of obscuring dust, and Valentijn estimates that, on average, 85 per cent of the stars in a typical spiral are hidden behind a dusty veil.

This means that we see only 15 per cent of the total starlight in a spiral galaxy; all the rest is concealed by dust, heaped up with clouds of molecular gas. There is another point, too. We know that the Sun, with its system of planets, lies about 40 light-years to the north of the main plane of the Galaxy— and yet we see more galaxies in a northward direction than in a southern. So, says Valentijn, there is more absorption of light in the southern galactic hemisphere than in the north, and this again must be due to dust.

All this could be more significant than might be thought at first. We can find out how the stars in outer galaxies move, and it seems that stars in the outskirts of spiral systems whirl round the centres of their galaxies much faster than they ought to do if they were being affected only by the gravitational

The Frozen Lake of Triton: Voyager 2, 25 August 1989. The flooded basin is 120 miles wide and 240 miles long. The vent from which the flood erupted seems to lie near the right-hand end of the basin. Reproduced by kind permission of NASA.

The Whirlpool Galaxy, as drawn by the third Earl of Rosse in 1845. He was the first to see the spiral form, using his great 72-in reflector. The drawing is remarkably accurate.

pulls of stars which we can see. Astronomers have long thought that there must be a great deal of invisible material, but its nature is unknown. Is it 'dark matter' of a type which we cannot possibly detect by present-day methods? This has been a popular theory, but if Valentijn is right the explanation is much more straightforward. The excess of mass is due merely to stars which we cannot see because they are hidden by dust.

The basic idea is not new, but it has previously been assumed that the dust takes the form of a smooth layer right through a galaxy rather than being clumped up in definite clouds. Probably we will find out in 1993, with the launch of ISO, the Infra-red Space Observatory satellite. Its instruments will be able to study the molecular hydrogen clouds in other galaxies, and this should tell us whether Valentijn's ideas are right or wrong.

97 Eggstraordinary!

It seems a long time ago since 1986, when Halley's Comet was in the news. Recently I was clearing out a drawer of newspaper cuttings of that period, and I found some which are decidedly amusing. There were the usual end-of-the-world prophets, of course, and there were also the commercial operators—reminiscent of the previous return in 1910, when an enterprising gentleman made a large sum of money by selling anti-comet pills (I never discovered what they were meant to do). And there was the Egg Episode. . . .

In 1910 it had been claimed that when the comet was at its closest and brightest (remembering that it was then much more spectacular than at any time in 1986), some hens laid eggs with comet patterns on them. The question was: Would the same thing happen again? In England, the Thames Valley Egg Company offered a prize of £10,000 for the first authentic Halley's Comet Egg of the year. The Managing Director, Mr David Watts, said that the egg 'would have to show a star with a tail etched in the natural pigmentation of the shell'. And sure enough, eggs started to be received.

The best was probably that of a hen belonging to a Mrs Franklin, of Warwickshire, which astounded even Inspector Kim Miller of the Ministry of Agriculture; he said 'I've seen eggs with every sort of imperfection, but never anything like this'. It certainly looked like a cometary marking, and I understand that the Thames Valley company offered it to a museum, though I never found out whether the museum accepted it—or whether the reward was paid.

I was ill-advised enough to mention the affair during a television broadcast, and, predictably, I started to receive pictures of eggs. One showed the impression of a comet on its shell, and the farmer concerned told me that 'the particular chicken which laid the egg was out all night'; it must have been having a good time! Another specimen from Rugby showed a patch with a long, cometlike tail.

(a)

> ## CHORUS GIRLS IN PANIC
> ### FEAR END OF THE WORLD
>
> *From Rock Island, Illinois, came this interesting dispatch;*
>
> FEAR HAS STRUCK THE MEMBERS OF A THEATRICAL TROUPE PLAYING HERE, RESULTING IN THE SUDDEN AND TEARFUL DEPARTURE FOR THEIR HOMES OF TWO OF THE CHORUS GIRLS. SIX OF THE GIRLS SUDDENLY DEVELOPED A FEAR THAT WEDNESDAY WAS TO BE THEIR LAST DAY ON EARTH, AND AS THEY FELT THAT THEY HAD NEGLECTED THEIR PARENTS IN TAKING TO THE ROAD WITH A SHOW TROUPE, THEY DESIRED TO MAKE AMENDS BY GOING BACK TO THEM TO FACE THE END OF THE WORLD. THEY SPENT AN HOUR CALLING UP THEIR PARENTS ON THE LONG DISTANCE TELEPHONE AND THERE WAS MUCH WEEPING AND WAILING. FOUR OF THE GIRLS WERE PREVAILED UPON TO OVERCOME THEIR FEARS AND REMAIN WITH THE SHOW, BUT THE OTHER TWO SECURED THEIR WAGES AND LEFT FOR HOME.

Halleymania! Two newspaper headlines: (a) 1910, (b) 1986. Human nature does not change.

When the total passed the fifty mark I began to feel rather overwhelmed, particularly when actual eggs started to arrive through the post. Finally, I did receive one specimen which was truly remarkable. It was a large, brown egg, and on its shell was a pattern which really did look like the head of a comet at the end of a long, curved tail.

I was suitably impressed, but unfortunately the egg did not qualify for a prize. It was laid not by a hen, but by a duck!

☾98 Fragments of Other Worlds?

Before the Space Age, meteorites represented the only material from beyond the Earth that scientists could actually examine. They are either stones, irons or a mixture, and it seems very probable that most of them originate in the asteroid belt. There may be no difference between a very small asteroid, only a few tens of feet across, and a huge meteorite such as that which is still lying where it fell at Hoba West, near Grootfontein in Southern Africa, whose weight is over 60 tons.

But there are a few of the 10,000 meteorites in official collections which seem to be rather different. There are only eight of them, and they are called the SNC Meteorites after the regions in which three of them were found: Shergotty in India, Nakhla in Egypt and Chassigny in France.

(b)

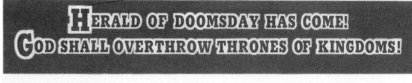

The SNC meteorites appear to consist of rocks which crystallized only about 1.3 thousand million years ago, as against an average age for meteorites of around 4½ thousand million years (much the same as that of the Earth). Their composition and texture indicate that they formed on or in a planet which had a strong gravitational field, and they have concentrations of volatile elements with glassy inclusions which were presumably formed in the extreme heat of whatever process ejected them from the parent body. These glassy intrusions contain trapped gases such as argon, krypton, xenon and nitrogen.

As long ago as 1979 some investigators suggested that the SNC meteorites were fragments from Mars. It is true that Mars has a strong gravitational pull (the escape velocity is over 3 miles per second) and that the texture and composition of the meteorites resemble that of the Martian soil, as revealed by the soft-landing Viking probes, but it is hard to see how the SNC meteorites could have been blasted off the surface of Mars by the impact of a larger body. They would have been vaporized.

John O'Keefe and Thomas Ahrens, of the United States, have suggested that a meteorite about a mile in diameter, striking the Martian surface at an angle of between 25 and 60 degrees, would vaporize large amounts of itself as well as the Martian crust, and the water and carbon dioxide vapour trapped in it, forming a jet of high-velocity gas which would sweep up pieces of the crust and blast them clear of Mars. They concluded that this is how the SNC meteorites found their way to the Earth.

It has also been proposed that five meteorites found in Antarctica come from space—not from Mars this time, but from the Moon. Antarctica has become a meteorite-hunter's paradise; many specimens have now been found, and have lain undisturbed for a very long time indeed. The Swiss geologist Otto Eugster has claimed that five of the Antarctic meteorites which he has

Meteor crater, Arizona. A rather unusual view, taken from a helicopter as we were making a television programme. If a meteorite capable of making a crater of this size landed in London, there would not be much left of the city!

examined are of lunar origin; his general idea is the same as O'Keefe's—a missile around 30 feet across hit the Moon and sent material away into space. He believes that his five Antarctic meteorites arrived on Earth around 75,000 years ago.

Whether or not either of these views is correct remains to be seen. Personally I doubt it, if only because I do not believe that the lunar and Martian craters are of impact origin; the non-random distribution argues against it, and it is also true that the lunar samples brought home by the Apollo astronauts and the Russian sample-and-return probes are very deficient in meteoritic material. But I admit that this is a minority view, and there is a distinct chance that some of these curious meteorites do indeed come from the Moon or Mars.

99 The Star in the 'Little Man'

Near the brilliant Canopus, unfortunately too far south to be seen from Britain, lies one of the most remarkable objects in the sky: Eta Carinæ, in the Keel of the now-dismembered constellation of the Ship Argo. During the 1840s it was the second brightest star in the whole of the sky; for a century now it has hovered on the fringe of naked-eye visibility, but it is quite unpredictable, and it remains very much of an enigma.

During the 17th century it was regarded as an ordinary star of about the fourth magnitude, but then it started to brighten up, reaching its peak in 1843, after which a steady decline set in until the brightness had fallen to about the sixth magnitude. Telescopically it does not look like a normal star; I have described it as 'an orange blob', and it is associated with nebulosity. The brighter inner part of the nebula, discovered about eighty years ago, has been nicknamed the 'homunculus' (Little Man) because it has appendages vaguely resembling a head and feet; it is actually an expanding dusty shell about two-thirds of a light-year in length.

We have long known that Eta Carinæ is exceptionally luminous. It seems to be at least 4,000,000 times as powerful as the Sun, and possibly more, so that it is one of the true searchlights of the Galaxy; its distance is 9000 light-years, and its spectrum does not fall into any particular class. Moreover, the spectrum is variable, and it is clear that Eta Carinæ is highly unstable.

It was once thought that what we now see is the remnant of a massive star which exploded in or about 1843. However, infra-red studies made in 1969 showed that it is still the brightest infra-red source in the sky apart from the Sun and the Moon, so that the star has not been destroyed; all that has happened is that it has hidden itself behind a dusty cloak of its own making.

This is where we can draw upon the power of the Hubble Space Telescope. Splendid images of Eta Carinæ have been obtained, showing that the homunculus nebula is clumpy, with clumps about ten times the size of the Solar System. In fact, the homunculus is probably a very thin and well-defined shell rather than a filled volume, either ejected from the star in a single burst or else swept up from surrounding material. A bright area to the south-west

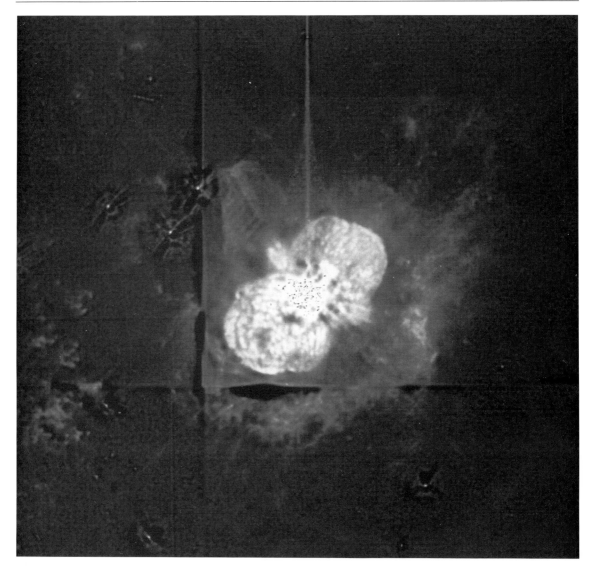

Eta Carinæ, as shown by the Hubble Space Telescope. The dusty nebula expelled from the star is probably a thin shell of material; ejecta from the star slams into the slower-moving gas to create a ridge of emission to the lower right. Reproduced by kind permission of ESA and NASA.

of the homunculus is due to material which was sent out from Eta Carinæ at a speed of around 3,000,000 miles per hour; as this ejecta slams into gas which had been ejected at an earlier epoch, shock-waves are produced, and these cause a glow. Another unexpected feature is a jet of material flowing outward, ending in a U-shaped feature which we can call a bow shock—much as a boat moving through water will produce a bow shock ahead of it. The ladder-like structure shown by the Hubble picture presumably represents some sort of wave phenomenon.

Significantly, Eta Carinæ appears to be about 100 times as massive as the Sun. This is exceptionally great—stellar masses have a much smaller range than sizes or luminosities—and there is evidence that it may be in the pre-supernova stage. If it does explode, then we will be treated to a glorious display of cosmic pyrotechnics. When this will happen we do not know; it may not be for a million years, but it could equally well be tomorrow. Of all the objects in our Galaxy, Eta Carinæ seems to be the most promising supernova candidate. Astronomers in general very much hope that its gala performance will not be too long delayed.

C100 Onward to Pluto

Far out in the chilly wastes of the Solar System moves Pluto, the ninth planet, discovered in 1930 by Clyde Tombaugh after a systematic hunt. It is a strange body. Initially it was believed to be larger than the Earth, but as methods of measurement improved the estimated diameter went down and down—someone facetiously suggested that if this trend continued, Pluto might eventually vanish altogether! The current value is 1444 miles, in which case Pluto is not only the smallest of the planets but is also smaller than several satellites, including our own Moon.

Another surprise came in 1977, when James Christy, at the US Naval Observatory in Arizona, discovered that Pluto is not a solitary traveller. It has an attendant, now named Charon, which is 753 miles across, and has therefore more than half the diameter of Pluto itself. We are not dealing with a conventional planet-and-satellite system; we are dealing with a double planet—or perhaps a double asteroid?

It had been known that Pluto takes 6 days 9 hours to spin once on its axis. This is also the revolution period of Charon, so that the two are 'locked' in an unique fashion. From one hemisphere of Pluto, Charon would always be visible, and would remain fixed in the sky; from the other hemisphere it would never be seen at all. And the distance between the two is a mere 18,000 miles.

Pluto's rotation period had been worked out from its periodical variations in light. In 1973 Leif Andersson and his colleague, J. D. Fix, found that the mean brightness of the planet was falling away, but that the variations due to spin were increasing. From this they deduced that there are bright poles, which were gradually being turned away from the Earth, and that there was a darker region near the equator—a remarkable result when we remember that Pluto looks like little more than a speck of light even in our most powerful telescopes.

The discovery of Charon provided a new opportunity. Andersson found that for a six-year interval every 248 years (the time which Pluto takes to complete one orbit round the Sun) Pluto and Charon undergo mutual occultations and transits. By a lucky chance this period fell in the 1980s, so that, for instance, Pluto could be seen 'on its own' while Charon was behind it. The opportunity was too good to be missed, and a long series of observations was made.

It has been found that the two bodies are not alike. Pluto's surface is coated with methane ice, Charon's with water ice; Pluto has an extensive though

Clyde Tombaugh. He discovered Pluto in 1930; I took this photograph in 1980. He is standing by the blink-comparator device that he actually used in the search.

thin atmosphere, Charon none. Pluto, with a density 2.03 times that of water, contributes about 80 per cent of the total light of the system under normal conditions, and Charon is presumably much less dense.

The cold is intense, and it seems likely that Pluto's atmosphere is variable—so that when the planet is near aphelion (its furthest point from the Sun) the atmosphere condenses on to the surface. There may be long periods when Pluto, like Charon, has no atmosphere at all. Moreover, Pluto's orbit is eccentric, and when at perihelion it is closer to the Sun than Neptune can ever be. Perihelion fell in 1989; not until 1999 will it regain its title of 'the outermost known planet'. There is no fear of collision, because Pluto's orbit is also tilted at the high angle of 17 degrees.

What is the real nature of Pluto? We cannot be sure. It is not a normal planet; in size and mass it is much the same as Triton, the largest satellite of Neptune, and there is even a theory that Pluto too used to be a Neptunian satellite which broke free and moved off independently—though the discovery of Charon seems to render this unlikely.

What we really need, of course, is a Pluto probe. Unfortunately none of the Pioneers or Voyagers could go anywhere near it, but technically it would be possible to launch a mission in the year 2001, and use the gravitational pull of Jupiter to speed it on its way. In one plan, the probe would 'hitch a ride' on board a space-craft which NASA has already scheduled to make a close study of the Sun; it would move out into the asteroid zone, come back to fly past the Earth in 2003 and then make for Jupiter. When near Jupiter, the Pluto probe would be separated, going on to rendezvous with its target in 2014, while the main space-craft would head back toward the Sun. In Method 2, the Pluto probe would be quite independent. Again the launch would be in 2001, the fly-by of Jupiter in 2003 and the Pluto encounter in 2014. A modest rocket, of the type known as Delta-2, could suffice; the cost would be about the same as that of a single Patriot missile used against Iraq during the Gulf War.

Unquestionably it would be a valuable scientific mission, and if it does reach Pluto in 2014 I hope to hear all about it, even though I will by then have reached the advanced age of ninety-one.

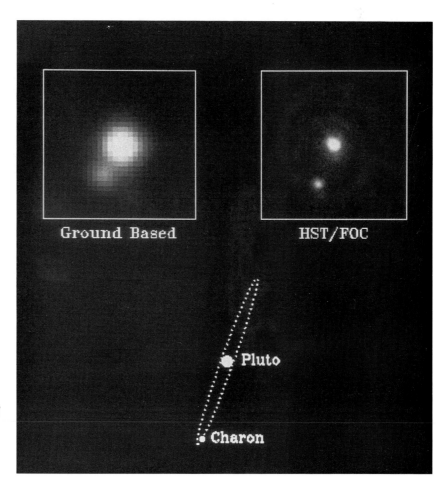

Pluto and Charon. The upper left picture shows the best ground-based picture of these two bodies—they are not clearly separated. The upper right picture was taken with the ESO Faint Object Camera on the Hubble Space Telescope, and shows the two clearly apart. Below we see the situation in the late 1980s, when Pluto and Charon showed mutual transits and occultations. Reproduced by kind permission of ESA and NASA.

☾101 What Does One Believe?

Human naïveté seems at times to be almost boundless. There are some ideas which appear to be so peculiar that they cannot be taken seriously—and yet they are. Of course, the classic case is that of astrology. There are millions of people who are quite convinced that the apparent position in the sky of a nearby planet, seen against the background of totally disconnected stars, can have a real effect upon character and destiny. All in all, this is rather less plausible than believing in the true existence of Father Christmas.

Records of spectacular events in the sky go back a very long way. In Ancient China, an eclipse of the Sun was thought to be due to a hungry dragon which was trying to eat the Sun. Today we know that a solar eclipse is caused by the passage of the Moon in front of the brilliant solar disk, and that a lunar eclipse happens when the Moon enters the cone of shadow cast by the Earth—so that its supply of direct sunlight is cut off, and the Moon turns a dim, often coppery colour before it emerges from the shadow.

Yet there are still people who are worried, and this was demonstrated in February 1990, when there was a total eclipse of the Moon which was well

Total Eclipse of the Sun, as I photographed it from Mexico on 11 July 1991.

seen from India. The problem is that Moslems are not supposed to look at the eclipsed Moon (I am not sure why), and Moslem temples are locked up. Unfortunately, however, a major religious procession was due at this time, involving a silver chariot and the divine Lord Subramaniam. Something had to be done.

Urgent discussions took place. The local religious leader, Guri Kanchi Sangarachariar, was called in, and confirmed that the procession would have to be re-scheduled. This was done, and notification was sent to all those who were preparing to wait in the road for the chariot to pass by so that it could be greeted by breaking coconuts on the ground. All went well until some tactless worshippers insisted on using portable radio sets to broadcast loud pop music, thereby causing considerable offence to the Waterfall Hilltop Temple Committee. The Administrator, Mr C. T. Ramasamy, said meaningly that suitable action would be taken.

There is in fact absolutely no danger in looking at the Moon, eclipsed or uneclipsed, because although it may dazzle the eye it sends us virtually no heat. This is emphatically not so for the Sun, and, as I have emphasized many times, even a quick glance at the Sun through a telescope or binoculars will probably result in permanent blindness. But this can be carried too far. During the 1960s there was a total solar eclipse, visible from parts of Europe, and I was carrying out a live commentary from the top of a mountain in Jugoslavia. Subsequently my BBC producer overheard a conversation between two old ladies in a bus queue: 'Yes. I watched that eclipse thing, and I did as the man said. I put on my dark glasses all the time I was looking at the TV screen.' I wonder what she saw?

A different sort of problem arose in 1969, in the city of Bradford, which has a very large Moslem immigrant population. You may remember that in July 1969 Neil Armstrong and Buzz Aldrin made the first manned landing on the Moon, in the lunar module of the space-craft Apollo 11. At that moment I was in a BBC television studio in London, carrying out a commentary; I had just come back from attending a last-minute conference in the United States, as I had been one of the accepted 'Moon-mappers'. On the following day my commentary was re-broadcast in a Bradford school as an educational lesson. Alas! it did not go down well. Earlier in the morning children had been to the local mosque, and had been told that landing on the Moon was impossible; the mullahs had said so, and it was against the teaching of the Qu'ran. So, said the children, the whole thing was nothing more than blatant American propaganda.

I'm not sure what they thought about me—I never asked!

102 Bombardment in AD 2100?

The chances that the Earth will be hit by a massive body from space, capable of flattening any city unlucky enough to be near the impact point, are not very great. The Solar System is a large place, and the Earth, relatively speaking,

is a small target. Yet the chances are not nil, and this has led to a theory due to four British astronomers (Victor Clube, David Asher, Bill Napier and Mark Bailey) which I very much hope is wrong.

We know of many asteroids which pass close to us; one has even been known to pass between the Earth and the Moon. Now and again there is a major strike, as happened in 1908 in the Tunguska area of Siberia, blowing pine-trees down for miles around. The British astronomers have looked at one particular stream of interplanetary objects which includes one comet, Encke's, as well as several small asteroids and four well-studied meteor streams. They believe that all these came from the break-up of a giant comet between 10,000 and 20,000 years ago, which Clube and Napier had previously referred to as 'the Cosmic Serpent'. The idea ties in well with the age of a meteorite from this stream which fell in Kentucky in the late 19th century; its age is between 7,000 and 25,000 years.

The analyses indicate that every three thousand years or so there are three or four occasions when the Earth passes through the stream. This means that for a century or so we are subjected to displays of cosmic fireworks. The fragments are not strong enough to survive high-velocity entry into the atmosphere, so that they do not produce craters; instead the effect is rather like having violent nuclear explosions in the atmosphere.

The British astronomers claim that we have been coming up to one of these dangerous periods for some time, so that the Tunguska event of 1908 was one warning of it; moreover, we know that in June 1975 a shower of meteorites hit the Moon, and were recorded by the instruments left on the lunar surface by the Apollo astronauts. If the predictions are right, the next bombardment will start in about a hundred years from now, so that our descendants of AD 2100 will have to be prepared for a lively time.

Support for these theories, it is said, comes from records left by old civilizations—the Romans, Greeks, Mayans, Incas, Chinese, Babylonians and so on; even the New Zealand Maoris and the aborigines of Australia. Missiles from the sky were, apparently, almost commonplace.

I admit to being unconvinced, and I do not believe that there will be any catastrophic bombardment toward the end of the coming century. But I may well be wrong—and, of course, I won't be there to see!

C103 A Backyard Quasar?

Old woodcut of a 'meteor storm': this dates from 1872—the meteors represented the débris of a dead comet (Biela's).

Now and then there are great revolutions in astronomical thought. The first was the discovery that the Earth is a globe rather than being flat. Next it was found that the Sun, not the Earth, is the centre of the Solar System. Then, in our own century, Edwin Hubble proved that the blurred objects then called 'spiral nebulæ' are galaxies, star-systems in their own right, so remote that their light takes millions, hundreds of millions or even thousands of millions of years to reach us. Are we on the verge of another revolution?

All our measurements of the distances of objects far beyond our local group of galaxies depend upon what is termed the spectral red shift. If an object is moving away from us, it will look slightly redder than it would otherwise do, and the amount of reddening is a key to its velocity—which in turn, is a key to its distance, because the further away a galaxy is, the faster it is receding; the entire universe is expanding. This, of course, is the well-known Doppler effect. The actual colour change is normally too slight to be noticed, but it shows up in the spectra of the galaxies. All the dark lines are moved over toward the long-wave or red end of the spectrum.

(There is a familiar, everyday proof of the Doppler effect. If you listen to an approaching ambulance whose horn is sounding, the note is higher-pitched than it will be after the vehicle has passed by and has begun to recede. During approach, more sound-waves per second are reaching your ear than would be the case otherwise, and the wavelength is effectively shortened; during recession fewer sound-waves per second reach you, and the wavelength seems to be lengthened. With light-waves, the Doppler effect affects the wavelength, and hence the colour, in just the same way.)

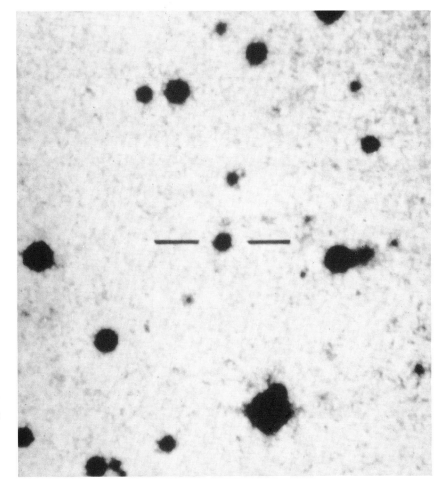

Quasar. This negative picture shows a quasar, 2000-330, indicated by the identification bars. It is generally believed to be immensely remote and super-luminous, but a few astronomers have their doubts!

In 1963 astronomers located some curious objects known first as quasi-stellar objects (QSOs) and now generally called quasars. They are generally believed to be the cores of very active galaxies, and their red shifts show that they are very remote indeed; some of them are well over 10,000 million light-years away, and are speeding away at over 90 per cent of the velocity of light. But—and this is the point!—everything depends upon the assumption that the spectral red shifts are pure Doppler effects. Just suppose that there is another factor to be taken into consideration?

This is the view of Dr Halton Arp, who spent many years in working at the Mount Wilson Observatory and who is one of the leading specialists in studies of galaxies. He has taken photographs of objects—galaxies and quasars—which seem to be associated with other, sometimes connected by bridges of luminous material, and yet have totally different red shifts. From this, he claims that the red shifts are not pure Doppler effects; there is a 'non-velocity' component, and this would mean that all our distance measurements need to be drastically revised.

Dr Halton C. Arp. He believes that the spectral red shift is unreliable as a method of distance-measuring.

Together with Geoffrey Burbidge of the University of California, Arp has recently been studying a quasar known as 3C-273. This appears as the brightest of all the quasars (you can see it, as a speck of light, with a modest telescope), and its spectrum indicates that it should be about 2,000 million light-years away. Yet Arp has found that it seems to be interacting with a giant elliptical cloud of hydrogen gas in the constellation of Virgo, whose distance is no more than about 65,000,000 light-years. If this is true, we have objects with very different red shifts which are nevertheless at the same distance from us. The implications are staggering. Modern cosmological theories would have to be thrown overboard; quasars, far from being super-remote and super-luminous, could be more or less in our cosmical backyard.

Arp has been saying this for a long time, and he has had some influential supporters, notably Sir Fred Hoyle. There are now many cases of objects which seem to be linked and yet have different red shifts. Conventional astronomers put this down to mere chance alignments, and dismiss the case of 3C-273 as 'an amusing coincidence'. But is it? Arp is one of the world's most skilled observers, and to disprove his theories means first disproving his evidence.

We await further results. All I can say at the moment is that if Arp is right, then most other people are wrong, and from the theoretical point of view we will be back if not in Square One, at least in Square Two. If so, we will indeed be in the throes of a gigantic upheaval.

104 Dead Cows in the Desert

About 3500 years ago a large meteorite, at least three feet across and with a mass of several tons, fell in the Atacama Desert of Northern Chile. During its descent it broke up into numerous smaller pieces, which impacted in the desert sand over an area of around eight square miles—where they remained, in an excellent state of preservation because of the very dry climate.

When prospectors travelled through the region in the 1860s, in search of valuable minerals, they found meteoritic fragments which they first thought to be of silver. They collected some of the fragments and brought them into the town of Copiapo—perhaps a thousand kilogrammes altogether—and many of them were lost, but 45 kg of the material found its way into mineral collections, and the meteoritic nature of the material was soon recognized. The meteorite fall was given the decidedly unattractive name of Vaca Muerta (Dead Cow) after a nearby dry riverbed, the Quebrada Vaca Muerta. Subsequently the location was forgotten. It was rediscovered in 1985 by Edmundo Martinez,

Fragments of the Vaca Muerta Meteorite. Harri Lindgren examines a minor's depôt of fragments from the Vaca Muerta Meteorite. Reproduced by kind permission of the European Southern Observatory.

who was then a geology student at the Universidad del Norte, Antofagasta. (He now runs a travel agency in San Pedro de Atacama, though whether he can manage a roaring trade seems to me to be rather problematical.)

Martinez discovered one large meteoritic mass, and this became known to the Chilean geologist Canut de Bon, who decided to investigate. Together with two astronomers from the European Southern Observatory, Harri Lindgren and Holger Pedersen, de Bon explored the site, and over a period of four years 77 specimens of the Vaca Muerta meteorite were collected, with a total mass of over 3400 kg.

The meteorite is of the rare stony-iron type known as a mesosiderite, and the new finds have more than tripled the amount of this material available for study. The distribution of the recovered fragments (the fall area measures 7 miles by 1¼ miles) indicates that the meteorite entered from the east–south-east, so that it flew over the Andes Mountains before impact. One of the largest fragments landed with such force that it made a crater 32 feet in diameter and almost 7 feet deep.

Twenty of the specimens had previously been collected by miners, and made into objects such as tools, cooking utensils, cans, corks, horseshoe nails, and in one case a coin. However, 57 specimens, ranging in mass from a few grammes to one piece weighing 309 kg, were in their original condition. All the material is now in Chilean collections, mainly the Universidad de La Serena and the Museo Nacional de Historia Natural in Santiago.

It was an exciting find, and one is bound to wonder how many more 'dead cows' may still be lying in the desert, even though it is likely that most of the surviving fragments have been collected. An apt comment came from Holger Pedersen: 'We are used to observing remote objects in space, but it really was great fun for once to do some down-to-earth astronomy!'

C105 A Surprise from Halley's Comet

Most people will remember Halley's Comet. It returns to perihelion every 76 years, and at some returns it has been brilliant; according to reports, it has even been known to cast shadows. Unfortunately its last appearance, in 1986, was as unfavourable as possible, because the Earth and the comet were in the wrong places at the wrong times. However, it was easily visible with the naked eye, and a whole armada of space-craft was sent to it. One probe, the European Giotto, penetrated the comet's head, and took close-range pictures of the nucleus. Subsequently the comet drew away, and faded, but large telescopes were still following it at the end of the decade. Most of the main work was then being carried out with the Danish 60-inch reflector at the observatory of La Silla, in Chile.

Halley's Comet has an icy nucleus shaped rather like an avocado pear, roughly 10 miles long by 4 miles broad. It has a dark coating, made of dark

Halley's Comet from 777,000,000 miles. This picture was obtained with the Danish 60-in telescope at La Silla, by combining 50 CCD frames taken over 19 nights in April and May 1988—before the mysterious outburst. The nucleus is of magnitude 23; it is surrounded by the asymmetric inner coma and by the much larger, elongated coma. Reproduced by kind permission of the European Southern Observatory.

The Ariane rocket, with Giotto. (Facing page.)

grains which are carbon-rich and therefore organic; they reflect only 4 per cent of the sunlight falling upon them. This was the first surprise when Giotto made its foray; up to that moment most astronomers (not all) had expected the nucleus to be bright.

As a comet approaches the Sun it is warmed, so that the ices on the crust and below the surface begin to sublimate (i.e. turn straight from a solid into a gas) and a cloud forms round the nucleus, while dust grains fall out and a coma soon appears. A tail or tails may then be formed, dust-tails by the pressure of sunlight and gas-tails by the fast particles making up the solar wind. When a comet is near the Sun it may produce outbursts from vents in the surface. When it moves outward again, it cools; everything freezes; no more gas or dust is lost, and the comet goes into hibernation. That is what Halley did. By 1988 only a thin cloud was left round the nucleus, and it was barely perceptible when I looked at it through the Danish telescope on 4 January 1989. By 1990 only the nucleus was left. The comet was then more than 1,200,000,000 miles from the Sun, between the orbits of Saturn and Uranus.

Suddenly, on 12 February 1991, Oliver Hainaut and Alain Smette, again using the Danish telescope, were staggered to find that their picture of the comet showed an extended 'cloud' three hundred times brighter than Halley

had been expected to be. After a few more nights there could be no doubt; Halley's Comet had undergone a major outburst. Since it had lasted for days, it was clearly being continually replenished from the nucleus, even though the temperature had fallen to about 200 degrees below zero Centigrade. It took weeks for the comet to fade back to its predicted brightness. So what had happened?

Three main suggestions have been put forward. First, the nucleus may have been hit by a wandering body, but this seems unlikely; we do not know how many such objects there are in the outer Solar System, but in any case one hit could hardly lead to a prolonged outflow. Secondly, there may have been the release of a large amount of energy stored in the nucleus—but why, after such a long time, unless for some reason or other a crack developed in the crust? The third suggestion involves interaction with particles of the solar wind, but it seems very doubtful whether these particles would carry enough energy to heat the surface of the comet sufficiently to produce such an outburst.

Frankly, we do not know the answer. Halley's Comet is still drawing away, and will continue to do so until the year 2023; what it will be like when it next returns, in 2061, is anybody's guess. I will not see it, but perhaps some of you now reading this book will still be around, so that you will be able to find out.

☾106 The Lunar Telescope

A century ago, the world's largest telescope was the strange 72-inch reflector home-made by the third Earl of Rosse, and set up at Birr Castle in Central Ireland. It had a metal mirror, and was mounted between two massive stone walls, so that it could swing for only a short distance to either side of the meridian; it was never clock-driven, and it could not be used for photography. Even so, Lord Rosse used it to discover the spiral forms of the objects we now know to be galaxies.

In 1917 came the Hooker reflector at Mount Wilson in California, which was for many years in a class of its own; Edwin Hubble used it to prove that Lord Rosse's spirals are huge systems containing thousands of millions of stars. Next in line, in 1948, was the Palomar 200-inch reflector—and with it Walter Baade was able to show that the universe is twice as large as anyone had expected. Today, pride of place so far as sheer size is concerned goes to the Keck Telescope atop Mauna Kea, in Hawaii, which has a 396-inch mirror made up of 36 hexagonal segments, fitted together to produce the correct optical curve.

Yet all these telescopes suffer from one defect: they have to operate from within the Earth's dirty, unsteady atmosphere. The Hubble Space Telescope, with its 94-inch mirror, is in orbit, so that 'seeing' conditions are perfect all the time—and despite its faulty optics, the HST has produced magnificent results. What will come next?

The 72-inch reflector; the Third Earl of Rosse is seen observing.

Astronomers are in no doubt that the ideal site for a major telescope is the surface of the Moon. There are many advantages. First, the Moon has no atmosphere, so that in this respect the conditions are as good as they are in space. Radiations in all parts of the electromagnetic spectrum will be available, whereas on Earth many of them are blocked by the screening effect of our air. Also, everything on the Moon has only one-sixth of its Earth weight, because the lunar gravity is so much less; this will make handling of 'heavy' materials much easier. Moreover, the Moon is a slow spinner, and an object will move relatively slowly across the lunar sky. The interval between sunrise and sunset will be almost two weeks, instead of a mere 24 hours, which is an extra bonus.

For a lunar telescope, astronomers envisage an instrument capable of working at wavelengths all the way through from ultra-violet to infra-red. In the middle of the long night, the Moon's surface temperature drops to around -173 degrees Centigrade, and this pleases the infra-red astronomers, because on Earth everything in the most sensitive part of the equipment has to be artificially cooled. The temperature of the lunar telescope could rise to 100 degrees Centigrade at noon, but even then it should still work.

Because of the low gravity, operating a large mirror on the Moon will be much easier than it is on Earth; there will be less strain and distortion when the telescope is pointed in different directions. We can safely plan apertures of 600 inches or more. The resolution should be amazing, and there is no reason why we should not be able to see planets of other stars (assuming that they exist, which seems very probable). Undoubtedly a lunar telescope would revolutionize our whole view of the universe.

Radio telescopes are also planned. On Earth there is increasing trouble with military and commercial interference. (Sir Bernard Lovell once said to me that unless action were taken, Earth-based radio astronomy might be confined entirely to the second half of the twentieth century.) On the far side of the

Moon, which is always turned away from us, there can be no interference at all; the environment will be completely radio-quiet.

When will all this happen? It is hard to say; so much depends upon military and political developments. But it is certainly not far-fetched, and we can only wait and hope for the best.

☾107 How to Welcome an Alien

Most astronomers, though not all, believe that life is likely to be widespread in the universe, and that there must be many civilizations living in other Solar Systems, many of them far more advanced than we can pretend to be. If so, then there is always a chance that the Earth will be visited. How would we react?

This was the subject of a special meeting held during the General Assembly of the International Astronomical Union, at Buenos Aires, on 26 July 1991. After discussion, a 'Declaration of Principles Concerning Activities Following the Detection of Extra-Terrestrial Intelligence' was adopted, and I think that it is well worth summarizing here.

The Declaration begins with the recognition that the search for extra-terrestrial intelligence (ETI) is 'an integral part of space exploration, and is being undertaken for peaceful purposes and for the common interest of all mankind'. Since any initial detection may be incomplete or ambiguous, 'it is essential to maintain the highest standards of scientific responsibility and credibility', and various principles of behaviour are laid down:

(1) Any individual or institute believing that any sign of ETI has been detected, should seek verification and confirmation before taking further action.

(2) Before making any such announcement, the discoverer should promptly notify all other observers or organizations which are parties to this Declaration. No public announcement should be made until the credibility of the report has been established. The discoverer should then inform his national authorities.

(3) After concluding that the discovery is credible, the discoverer should inform the Central Bureau for Astronomical Telegrams of the International Astronomical Union, and also the Secretary-General of the United Nations. Other organizations to be notified should include the Institute of Space Law, the International Telecommuncation Union, and Commission 51 of the International Astronomical Union.

(4) A confirmed detection of ETI should be disseminated promptly, openly and widely through the mass media.

(5) All data necessary for confirmation of detection should be made available to the international scientific community.

(6) All data relating to the discovery should be recorded, and stored permanently in a form which will make it available for further analysis.

(7) If the evidence of detection is in the form of electromagnetic signals, the parties to this Declaration should seek international agreement to protect

A Martian comes to London! How do you say 'Good morning'?

An alien in a BBC studio. (Do you recognize it?)

the appropriate frequencies. Immediate notice should be sent to the Secretary-General of the International Telecommunication Union in Geneva.

(8) No response to a signal or other evidence of ETI should be sent until appropriate international consultations have taken place.

(9) The SETI (Search for Extra-Terrestrial Intelligence) Committee of the International Academy of Astronautics, in coordination with Commission 51 of the International Astronomical Union, will conduct a continuing review of all procedures relating to the detection of ETI and the subsequent handling of the data.

Such a Declaration would have seemed quite unnecessary only a few years ago; times have changed. But when will ETI actually be detected—if ever?

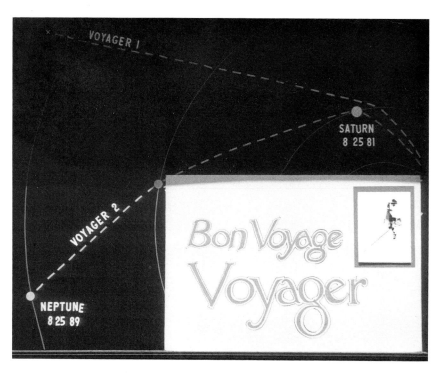

Goodbye, Voyager 2. This board was displayed at the JPL (Jet Propulsion Laboratory) after the Voyager 2 pass of Neptune in 1989. Voyager will never return; it carries a record to indicate its place of origin should an alien civilization ever find it. If they do detect it, what will they think, I wonder?

Index

DATE DUE			
GAYLORD			PRINTED IN U.S.A.